Reproduction
and Man

R. J. HARRISON M.A., M.D., D.SC.
Professor of Anatomy, London Hospital Medical College

Reproduction and Man

The Norton Library

W · W · NORTON & COMPANY · INC ·

NEW YORK

Preface

Quite apart from academic considerations there are many reasons why we need to know all we can about reproductive phenomena. The rapid increase in world population brings dangers to man's well-being which are not limited to those associated with insufficient foodstuffs, dietary deficiencies, famine and disease. Too many people means too little space, too few opportunities, crowded facilities, increased leisure; all affecting the full development of human personality, of humanism. Family planning, contraception and the 'pill', eugenics, abortion, are matters now openly discussed and on which both medical men and lay people have to make decisions that affect themselves, their patients, their families, and even the structure and behaviour of society.

This short volume selects and describes the main facts behind the 'facts of life' and discusses their significance to man. It is only natural that it includes much information gleaned from mammalian reproductive patterns. It was their investigation that historically gave the clues to understanding what happens in human reproduction. Aristotle, Pliny, de Graaf, Harvey, von Baer and F. H. A. Marshall all devoted much of their time to learning what reproduction in animals could tell them. Their successors have followed their example. Yet, despite the activities of the thousands of anatomists, physiologists, endocrinologists and biochemists who study reproductive phenomena, there are still many creatures that reproduce in ways not yet fully understood. F. H. A. Marshall began his classic work *The physiology of reproduction* with a quotation from Ecclesiastes: 'To every thing there is a season, and a time to every purpose under the heaven.' Many consider it to be

already late enough to put to practical use information which
may well conserve the human species.

Grateful thanks are extended to Mr Frank Price, Mr R. F.
Birchenough and Mr D. A. McBrearty, for artistic and technical assistance; to Professor W. J. Hamilton, Dr. Ernest Neal
and the late Mr Alan Brews, with whom the author has calla-
borated for many years on problems reproductive; and to Miss
B. J. Fuller, for frank criticism, and to Miss E. J. Oliver, who
helped to prepare this volume. Mr I. A. G. Le Bek, Editor of
Contemporary Science Paperbacks, gave much assistance and
helpful advice.

R. J. HARRISON

Contents

1. Introduction

'I could be content that we might procreate like trees, without
conjunction, or that there were any way to perpetuate the
world without this trivial and vulgar way of union: it is the
foolishest act a wise man commits in all his life; nor is there
anything that will more deject his cool'd imagination, when he
shall consider what an odd and unworthy piece of folly he hath
committed.'

Religio Medici SIR THOMAS BROWNE, 1643

That man must reproduce to replace himself and that men
and women have a sacred and undeniable right to procure
progeny have been secular and religious tenets for thousands
of years. That he might reproduce too often and too rapidly
were dangers our ancestors scarcely considered. Rather has
the emphasis been to encourage parenthood. The enjoyment of
children, the pleasures of family life, child allowances from the
State, the social stigma of childlessness, the marriage service,
even festivities – what would they be without the children? –
all advocate the advantages of reproduction. People the
world, was the cry; now we have.

When numbers of men were sparse and widely distributed
the dangers of over-population were hardly apparent, and
when they occurred it was usually far from civilisation, or
impinged only occasionally. Famine, disease and war could
have other explanations as to their cause: idleness, ignorance
and ambition, or so it could be argued. Today, however, the
pressures of over-population are being exerted not only where
men are dying from famine but also where, surprisingly, they
are succumbing to overeating, obesity and affluence.

Sir Charles Darwin, a grandson of the Charles Darwin,
wrote recently that we were living in 'an entirely abnormal

1

period of history'. One of the principal abnormalities which
he emphasised was that the world population was becoming
doubled during a century. He dramatically illustrated one
result of the doubling world population each successive
century by calculating that by about A.D. 3954 there would be
so many people that there would be standing room only for
them on the Earth's surface. Such a result is impossible, but
it is undeniable that the rate at which the world population is
now increasing only allows this absurd, yet disturbing conclusion. At the *present* rate of reproduction the world population is increasing by some 36 million a year, and all indications
are that this rate of increase will accelerate. From a level
somewhere about 470 million in 1650 the population of the
world is estimated to have risen with increasing rapidity to
1171 million in 1850 and to nearly 2500 million in 1950. From
3000 million in 1960 it is estimated that world population is
now likely to *double* in only the next 40 years.

Table 1. *Estimates of world population at various dates.*

Year	Estimated total *in millions*	Estimated increase
1650	470	
1750	728	55% in 100 years
1800	906	
1850	1171	62% in 100 years
1900	1608	
1950	2377	103% in 100 years
1961	3000	
1975	4000	
2000	6000	100% in *40* years

The rapid increase in world population is already causing
concern to many economists, agriculturists, scientists, and to
Governments, particularly that of India. An inescapable conclusion is that every effort should be made to reduce the rate of
increase and, if possible, to stabilise world population at a
reasonably early date. Otherwise, the steadily increasing
world population will rapidly use up the available material and
nutritional resources. These gloomy prognostications had been

made earlier by the kindly Surrey clergyman and teacher, T. R. Malthus, in his *Essay on the principle of population* published in 1798. He foresaw the population eventually and inevitably consuming the available food, and it is astonishing how much fury his arguments engendered. Granted, he was not always consistent in his views, but they sent Karl Marx into a frenzy of vituperation, echoed by lesser voices. Not that Malthus had only original views, nor, indeed, did he put forward any single all-embracing hypothesis. He was also rather too inclined to moralise, which, as well as infuriating his readers, took the edge off his concepts of effective demand and under-consumption in economics. Lord Keynes did much to revive regard for Malthus, even against the fears in the 1930s that there was a real danger of a population *decline*. Some recent writers such as Sir John Russell, Harrison Brown and René Dubos have been more optimistic and point out that man has still to take advantage of the technical possibilities of obtaining food from under-developed countries, from the sea, air and other sources. Recent estimates suggest that the widespread use of modern agricultural methods, together with better extraction of concentrated protein from plants, better use of solar energy in photosynthesis, and introduction of chemical methods for making synthetic foods, could make available enough nutriments to support a population not just twice the size of the present one but 20 times the number. Such techniques may well result in the traditional square meal looking rather different on a plate, but man has seldom been slow to adapt to new eating habits. Even so, the implications of Sir Charles Darwin's mathematics are still, at the very least, cause for reflection.

The problems associated with increasing population not only concern consumption of food and control of disease but also problems of birth control, family size and old age. It seems somewhat perplexing to find that the main contributions of medicine during the last century have been an increase in the live birth-rate and a prolongation of the expectation of life. In general, it has been medical progress that has ensured the birth of, and prolonged the life of, those members of the population who in earlier centuries would have died at birth or lived a shorter life. It can be argued that this represents a reversal of the law of the survival of the fittest and has even developed a cult of looking after the unfit. The advances brought about by improved medical attention would be more advantageous for the individual if the population could be stabilised. Medicine has indeed set major problems by solving many of the more immediate ones.

It would seem that by having learnt to combat his natural enemies, pathogenic micro-organisms and parasites, man is doomed to overrun the Earth with his multiplying numbers as long as famine, accidents and war do not take a significant toll. This seems inevitable as long as man does not interfere with his reproductive processes, control his fecundity or become influenced by other demographic forces. Many factors control population size* some are biological, some social or economic, and some are the 'vicious practices' condemned by Malthus. Recently it has been suggested that there are factors governing mating or marriage that will tend to stabilise populations by an intrinsic influence. The concept of varying nubility in women suggests that the more desirable qualities, which make for material success in life, are likely to become genetically associated with lower fecundity.

Success in life, in its many and varied forms, is naturally the driving ambition of every unit of reproduction, of every individual of a species. Over-population can have dramatic effects on the very nature of this success, and for man his world is being ruthlessly changed by a series of social upheavals and inventions which continually modify the rewards available. From the time of his discovery of fire and the invention of the wheel, to the industrial and social revolutions of more recent times, the associated and steady increase of his numbers has changed man's ways. Undoubtedly these factors have been advantageous, but an increasing, thriving, human population can hardly be expected not to exert mounting internal pressures in the social environment. The nature of these pressures,

* See the *Report of the Royal Commission on Population*, H.M.S.O., 1949.

which Malthus could hardly have foreseen, and how to circumvent them, must be the concern of all of us. They are causing a form of creeping paralysis of man within an environment which is gradually becoming impoverished. Let us call them menaces of manufacture. They are: (a) too great a dependence on electricity; (b) over-motorisation; (c) purposeless distraction by entertainment machines; (d) the superiority of the computer over the abacus. They will lead to struggles to find space in which to live and frustrations as to what to do with leisure which cannot be purposefully enjoyed. Affluence is said to breed influence; in the future it seems more likely to lead to idleness, sloth and atrophy.

The loss of freedom within the available living space is exerting other, little-considered effects, all attributable to over-population. There is destruction of wild-life, both animal and vegetable. This removes man from natural surroundings and from contact with nature. Instead, he spends his life surrounded by buildings, factories, roads, railways, all of similar materials and too often depressingly alike. Privacy, quiet and solitude, meditation and fantasy, all necessary to man for his very humanity, are becoming more difficult to attain. Instead, he has to become socially integrated, to conform in his 'togetherness', until he is really little better than the rodents he studies in his laboratories – caged, labelled, massed and all alike.

Malthus's great message was a warning about what we now call population 'explosion'. This means, of course, an overwhelming increase in numbers, but it would be better to talk of a population surge or avalanche and a surge in population density. Conversely, a sudden fall in numbers is called population 'crash'. These occurrences are by no means limited to human populations; they occur in groups as widely diverse as locusts and lemmings, where sudden increases in density are followed by severe declines. Other creatures exhibit various kinds of self-regulation in their numbers, which may be grouped into those that result in increased mortality and those that diminish reproduction. The former include suppression of normal defence mechanisms and the appearance of certain diseases that may result from infective organisms or from endocrine stress. Inhibition of reproduction results from the actions of such factors, and they also affect fertility (ability to produce young). There is almost always an associated increase in death of embryos while in the uterus, in death at birth, in the occurrence of abnormalities leading to sterility and of inadequate parental care.

There are, in fact, three main components of the growth in a population: natality, mortality and mobility. Natality means the number of young born as a function of the number of males and females, and mortality refers to the death rate. Mobility is essential where, as the result of local extinction or increase in density, individuals move in or out to repopulate or emigrate. Each species exhibits a geographical distribution, living most successfully in an environment and with other creatures (the whole complex called an eco-system) that suit its reproductive capacity. Within the area of distribution there will be local groups whose success or failure will affect reproduction. Should success occur all over the area, then there follows danger of impoverishment of the eco-system, commencing in those local areas that began to thrive first. Scientists are accused of going down brain drains for personal rewards. Perhaps they are exhibiting a response to the natural urge to leave impoverished eco-systems!

Most of the factors in an eco-system that operate adversely on growth of population are not difficult to understand. Clearly, food shortages, disease, predators, problems of dwelling space, floods, earthquakes and war, increase mortality, but some of these factors may also increase the subsequent natality. They stimulate competition for the available elements in the substrate on which the population grows, and at the same time they may provide a damping effect on the reproductive rate. More subtle controlling mechanisms may operate through the construction of hierarchies, or 'peck orders', or elaborate harems with a dominant bull whose bite endows it with seigniorial rights. These social conventions are frequently associated with territory establishment, breeding grounds, segregation of juveniles and thus the construction of blocks to promiscuity. Should a species exhibit migration, there may be a reduction in the time during any year when mating is possible; and should this be linked with polygamy, then over-crowding of the breeding ground could mean that only a proportion of females are served. It is not difficult to transfer most of such concepts to human reproductive behaviour.

More recently it has been discovered that various external factors can affect the internal complex of endocrine glands (pituitary, adrenal, gonad) concerned in controlling repro-duction. These 'exterocrinological' effects can be induced by factors such as the size of nesting chambers, the number of males in a limited space, the overall density of the population, the introduction of a strange male at the time of mating. It is as if behavioural and emotional responses to situations detrimental to reproduction cause a feedback and so lower natality. In man these responses could lead to wider acceptance of birth control, abortion and adoption. They could lead to a stricter morality or, conversely, a different view of marriage. There could even be, though we think it unlikely, the develop-ment of an attitude of shame at having large families – a tax on the fourth child, say. 'Gluttony,' read great-grandmother's sampler, 'is a deadly sin.' But she over-indulged in natality thirteenfold!

Natality in man, therefore, is more than just the expression of birth-rate. It is the termination of a complex series of genetic, anatomical, endocrinological, embryological and bio-chemical events, and has behavioural, social and economic backgrounds that are becoming increasingly interwoven. Interference with the natural natality may have unknown and unsuspected results. Does birth control of any type affect the vitality of a population? Are small families really desirable? How will official sanctions affect the population structure? Will the widespread use of a Western 'pill' alter the outcome of future war games?

We need to know all we can about the biology of reproduc-tion, not just in man, but in animals, especially domestic ones and those which might be farmed for the first time to increase food production. The deliberate use of game reserves in national parks in Africa to farm hippopotamus and elephant could be extended to other parts of the world. Several countries have been for many years dependent for food, and even clothes and drugs, from animals in the sea. The establishment of food chains artificially in seas of limited size by seeding with chemicals; introduction of selected species of plants and animals, and use of underwater farming methods may have to be explored with just as much enthusiasm as is accorded to reaching the moon. Just as we require knowledge of reproductive patterns in order to control growth of human population, so we also need it to know how to breed better and more abundant farm animals to eat.

The investigation of reproductive phenomena is really concerned with five main problems. They are: (i) the formation of germ cells (ova and spermatozoa); (ii) factors concerned in germ cell maturation, viability and release; (iii) transport and union of germ cells; (iv) establishment of early pregnancy and maintenance of normal development; (v) birth of the young and their immediate survival. These comprise a series of sequential events, all complex and delicately balanced. To interfere at any point in the sequence in order to control fertility may have profound consequences not obviously apparent.

Early methods of family limitation and population control were effective but somewhat unethical. The commonest were infanticide, starvation, abortion, barriers to early marriage, and ritual abstention from intercourse. Use of methods of contraception is older than we realise. There are references in Egyptian papyri to contraception dating back to 1850 B.C., and in ancient Hebrew literature and in Aristotle's works. There was little contraceptive knowledge in medieval Europe, but the sheath is mentioned in 16th and 17th century writings and was on sale in London in the late 18th century. Malthus did not advocate birth control, but he did plead for moral restraint; in his day this really meant late marriage. Francis Place was probably the first to recommend widely contraception in Great Britain in his *Some illustrations of the principles of population* (1822). Subsequently there was increased interest as the result of the activities of Charles Knowlton (*Fruits of Philosophy*), Charles Bradlaugh and Mrs Annie Besant, and after 1877 the topic of family limitation was widely discussed. Not until there were improvements in contraceptive methods, however, was there any effect on the birth-rate.

The invention of the vulcanising process of rubber (1844) may have had an important influence, but equally so did the growing disinclination of women to have large families and their wish to participate, less encumbered, in social and national life. Mechanical, irrigation, and other methods all vary in their acceptability and effectiveness with the temperament, care and intelligence of the user. Of the more modern methods, two emerge as being highly effective: oral contraception (the 'pill'), and the use of intra-uterine devices in the form of stainless steel, silver and plastic rings, spirals and loops. Both methods have now been on trial in several countries over a period of years, and the experience gained has shown that their advantages far exceed the disadvantages. Seldom are there definite contra-indications to their use by normal women. The increasing availability and use of oral contraceptives has raised a problem not yet widely grasped or clearly understood, despite the fact that in the United States the majority of young women with college training have already used oral contraceptives.* Their use exploits a new principle in the control of fertility, in that it separates the act of intercourse from the process of procreation. By a combination of two hormones, the liberation of an egg from the ovary at ovulation is inhibited. The ovary and the functions it controls are, generally speaking, 'put out of action' until the individual ceases to take the hormones, when a reduced menstrual bleeding occurs. This means that a woman using oral contraceptives is not really in possession of her full biological and thus reproductive faculties. The risk of pregnancy is removed, and the purpose of intercourse *could* be considered to have become altogether different. In another way, of course, a woman does in fact have control; it is for her to decide *when* her reproductive potential is available. Whatever the moralities involved in

* *Science*, 1966, vol. 153. p. 1204.

accepting oral contraceptives, there remain several definite
results. The disasters of unwanted children, back-street
abortions, unending childbearing can be eliminated. Whether
women will now become mothers so much later in their lives
that the birth-rate will fall, and families be smaller, remains to
be seen.

2. Patterns of Reproduction

Animals, and plants, live for a certain length of time and then die. They must be replaced if any type of organism is to survive on the planet, and the primary importance of reproduction is the making of a new generation of similar organisms to replace the one that dies. The number in the new generation is also important. A one-for-one replacement might be just enough to keep a species in existence provided there were no diseases, no failures and no predators. A replacement by many for one would certainly allow for losses, but might result in the number of a species multiplying to be exceedingly great – so great, perhaps, that the numerous individuals would eat up all available food and then the species would almost entirely die out due to starvation. Reproduction is necessary for survival of a species, but uncontrolled reproduction can be dangerous.

A second factor of great importance in the replacement process is that it produces occasional offspring which are slightly different in some way from their parents. The difference may be expressed in an anatomical feature, or an ability, either of which is of advantage to the new form in its environment. It prospers, as do those of its offspring who in turn inherit the new feature. The feature becomes selected. The mechanism producing the differences is that which brings about organic evolution. Reproduction is necessary for the unfolding of evolutionary change.

The simplest form of reproduction is exhibited by simple, unicellular organisms. Each divides into two. This is binary fission, and since no other elements are involved it is also referred to as an asexual form of reproduction. Another example is the budding off from the parent organisms of a

11

small portion which grows into a similar form; spore formation is yet a further example. Sexual reproduction essentially involves the union of two cells and the subsequent division of the united pair into a new individual. Some single-celled organisms show an inclination to unite one with another (conjugation), either temporarily or permanently, before multiplying. More complex animals (Metazoa) have two kinds of cell which are made in special organs and extruded from the parent. The special organs are the *gonads* (testes and ovaries), and the two kinds of cell extruded are the *gametes* (spermatozoa and ova). One gamete unites with one of the other kind (fertilisation) and forms a zygote. The zygote may be formed outside the parents, in the sea; it may originate in one parent and then be extruded in a special protective covering like a shell; or it may stay inside the parent. The formation of a zygote marks the true beginning of a new individual: at first consisting of only one cell, it divides into two, four, eight, sixteen cells and so on until a new adult is formed. This process is called ontogeny, meaning the development of a whole individual.

All organisms, even the simplest, exhibit a programme that unfolds during their lives. Single-celled organisms have a period of youth just after their formation during which they are usually most vigorous. If living in the right circumstances, it is the time when they most frequently multiply by binary fission. Later, changes occur within their cells which could be described as a form of maturity; at this stage some individuals conjugate to reproduce. Later still they exhibit an ageing process, and then die. Complex organisms develop from zygotes and pass through an embryonic phase to form *young* individuals. The young grow to a stage when their gonads become active producers of gametes (*puberty*). Growth continues until the organism has reached its proper size and is physically and sexually *mature*. It no longer displays any further growth changes, but runs out its allotted life span, more or less gradually exhibiting age changes (*senescence*) until it becomes effete and *dies*. It is characteristic of living matter for it to consist of discontinuous units (organisms) which each reach a particular size and shape (characters of the species) and live for a certain average time, the mean specific longevity.

Mammals, including man, exhibit this life cycle but with some interesting modifications. A definition of a mammal includes the characteristic that the young develop in a womb, or uterus. Mammals are, therefore, viviparous, retaining the zygote within the female until it develops into a young individual which is then extruded by a birth process. The young can then exist independently, but still need to receive nutrition for varying periods from the maternal mammary glands. The female mammal *suckles* her young, the young *suck* from the female breast. Newborn mammals are sexually and physically immature. Intra-uterine gestation means the young have to be small at birth, as there are obvious spacial limitations in a uterus. There is a period of adolescence of varying length, with puberty and sexual maturity usually developing before physical growth has ceased. Sexual activity in mammals is not by any means continuous from puberty onwards. Most mammals display a period in the year when the gonads of both sexes are fully active. This is the *breeding season*. The gonads are inactive or quiescent out of the season, and it needs experimental stimulation or removal to a different climate to produce any functional response.

EVENTS IN MAN'S REPRODUCTIVE LIFE

Man reaches *puberty* at the end of childhood, one-sixth to one-seventh of the way through the normally expected life span. Man's childhood is longer than the comparable period in any other mammal. The most important and dramatic changes occur in his nervous system during childhood. Neither his brain nor spinal cord are well developed at birth, and it takes many years for this tissue to mature. During childhood man learns to walk (at about one year of age), starts to talk (at about two years) and develops a growing control over his muscular system. At puberty the female reproductive organs start to pass through a succession of *sexual cycles*, each lasting about 28 days. The beginning of each cycle is marked by the phenomenon of *menstruation* (or blood loss); *ovulation* (the shedding of an egg from the ovary) usually occurs near the middle of the cycle. There is no annual breeding season in either sex; in other words, man can breed all the year round. Active changes also occur in the male reproductive organs at puberty and result in the ability to ejaculate *seminal fluid* containing spermatozoa. There is no sexual cycle in the male; the reproductive organs are continuously active. Puberty is followed by *adolescence*, a period of becoming adult, or *mature*: it is particularly drawn out in man, more so in the male. During adolescence numerous anatomical changes occur in both sexes (some start even before puberty) and each sex acquires *secondary sexual characteristics* of proportion, shape, distribution of hair and quality of voice. The female has a *menopause*, or cessation of reproductive function (the 'change of life'), at a modal age of 49, and is the only mammal so far known to live long enough to display this phenomenon. There is no change equivalent to a menopause in the male; reproductive ability subsides more or less rapidly as age advances. The gestation period, or duration of pregnancy, averages 267 days, but is more variable than in any other mammal for which records have been kept. The usual *number* of young born at a time is one, *multiple births* occasionally occur and *lactation* frequently lasts for nine months.

Puberty

The onset of puberty, or the menarche, occurs in man between the ages of 12 and 15, and at an average age of 13·5 years in the northern hemisphere. It is frequently maintained that puberty occurs earlier in warm climates. This has not been proved to be due only to temperature; other factors, such as the level and type of nutrition, are involved. The onset of puberty is gradual in man and in the higher Primates. Hormones called gonadotrophins (experimentally shown to be the cause of the onset of puberty) are produced by the pituitary gland in steadily increasing amounts until they cause gonadal activity sufficient to promote growth of hair, mammary tissue and changes in the skeleton (and sexual skin in some other Primates). Such pituitary activity often starts some years before menstruation, or even the first ovulation. The onset of puberty is thus not to be considered as a point in time, although the occurrence of the first menstrual discharge (menarche) is conveniently taken as indicating its onset, but to be a period during which the gonads gradually attain an overtly functional stage, and at which reproduction becomes potentially possible.

Adolescence and sexual maturity

It must be clearly understood that puberty and sexual maturity are not the same. Gonadal development is gradual; menstruation may at first occur at quite long intervals, and ovulation may be delayed until the menstrual cycle is well established. The period of time during which the gonads steadily develop and the secondary sexual characteristics become increasingly more pronounced is known as adolescence. The processes of reproduction and physical development culminate in sexual and physical maturity; the first stage is reached before the second. Sexual maturity implies that the individual has reached a stage when it is best equipped for reproduction. It is reached much later than puberty, although there is considerable individual variation. An animal is sexually mature, therefore, when its reproductive organs have attained a size and degree of activity that indicates full reproductive capacity, both as regards begetting, bearing and rearing young. Thus many consider that sexual maturity in women also involves the full development of a responsible maternal outlook and of psychological maturity. A woman may reach her full reproductive capacity from about her 16th to 22nd year; in the male it is a more difficult stage to assess, but in general sexual maturity is reached later than in women.

There is evidence that in female mammals there is a short period of 'adolescent sterility' during which ovulation is irregular or does not occur. Pregnancy during this period is potentially dangerous to both mother and young. Most mammals that usually bear many offspring at a time have fewer when they are younger; thus yearling ewe-lambs usually have only one offspring at their first breeding; as they grow older multiple births are more frequent. It does not follow that older women are more likely to have twins, but it is an illustration of the gradual attainment of sexual maturity.

Physical maturity assumes that the individual has reached a stage of full adulthood in all anatomical aspects. It coincides with the end of bone growth; it is reached earlier in women (20–24 years) than in men (23–26 years).

The female sex cycle

The first menstrual loss at puberty is followed by a succession of regularly repeated cycles, each cycle involving changes in the ovary and uterus. We may, therefore, refer to an *ovarian cycle* in the ovary and a *menstrual cycle* in the uterus. The menstrual cycle is under the control of the ovary; when the ovaries are removed menstruation ceases. The ovarian cycle is controlled by the anterior lobe of the pituitary gland and is initiated only when sufficient gonadotrophic hormones are secreted from the gland; other factors, such as the nutritional state, may have an influence on the time of the onset. Once started, the cycle recurs regularly, is controlled by the anterior pituitary, and continues, except when interrupted by pregnancy and lactation, until the menopause. Additional factors, such as adrenal and thyroid secretions, probably also affect the cycle. Other organs and tissues are also affected by the hormones controlling the menstrual cycle, but in none are the changes as dramatic as those in the uterus.

The phenomenon of menstruation in the human female is caused by the breakdown of most of the inner lining (*endometrium*) of the uterus and its loss, together with some blood, from the uterus to the outside through the vagina as the menstrual flow (*menses*). The first day of the flow occurs 21 to 34 days after the first day of the *last* menstrual period and a similar interval before the first day of the next expected period, as long as pregnancy does not intervene. The average length of one cycle is 28 days, and the average length of menstruation is five days, but only a small percentage of women have a cycle that displays such exactitude repetitively. Considerable variation is shown and is quite normal.

Menstruation ceases on the supervention of pregnancy and

does not recommence until some months after the birth of the child, depending on the length of lactation. If the child is suckled for a long period, this may well suppress the re-establishment of menstruation until that much later, but lactation does not necessarily suppress ovarian function all the time. So it is quite possible for ovulation to occur towards the end of the lactation period, although menstruation has not been re-established. If the child is not suckled, menstruation is usually re-established correspondingly earlier.

The most important event in the female sex cycle is the release of the maturing egg from the ovary. The entire cycle, both its ovarian and uterine components, is directed to ensuring the successful liberation of the maturing egg and, equally important, the success of the pregnancy should the egg become fertilised.

Ovulation in women occurs midway between two menstrual periods: that is, on the 13th or 14th day of a regular 28-day cycle. Until ovulation the maturing egg is in the ovary, contained at first in a small capsule of cells which later enlarges into a fluid-filled sac. The rupture of the sac, or ovarian follicle, is spontaneous in woman, as it is brought about primarily by hormonal influences and not by the stimulus of coitus. A few women are aware of slight abdominal pain about the time they ovulate. Usually only one follicle ruptures during each cycle, rarely two, and very rarely three.

The menopause

Menstruation ceases at the menopause, which marks the ending of the period of reproductive life in women. Essentially, the menopause must be considered a reflection of a gradual cessation of control by the pituitary over the ovary. Usually the menopause extends over several years, and is accompanied by signs and symptoms that can nearly always be referred to the cessation of action of ovarian hormones. In that it also marks the loss of an attribute – that of reproductive power – it is a stage in ageing of the human female organism. At first sight it may seem curiously human to have a menopause – for it is not known in any other mammal in the wild, though it is said to have occurred in certain Primates in captivity and occurs in modified form in laboratory rats. It may well have been rare in primitive man with his relatively short expectation of life, but its universal occurrence nowadays indicates that modern woman can outlive her period of reproductive life by many years. The present average expectation of life for a newborn female child is about 73 years, and with a menopause occurring at a modal age of 49 this means an average expectation of nearly a quarter of a century as an individual incapable of reproducing, and for that reason as an incomplete individual. This incompleteness does not mean that post-menopausal women are unable to make important and vital contributions to society, but it does raise certain individual problems, perhaps only in that women also have a slightly greater expectation of life than their husbands.

Pregnancy – the gestation period

A woman is potentially pregnant, or has started her conception, when the ovulated egg is fertilised by a spermatozoon. We must write 'potentially' pregnant, because not every fertilised egg, or zygote, inevitably develops into a live baby: there is a loss at certain stages in development. We refer to the *products* of fertilisation generally as a conceptus.

The egg is fertilised shortly after ovulation and takes some five to six days to reach the uterus. It is travelling from the ovary to the uterus along the uterine tube and developing slowly during the journey. The uterus is simultaneously being prepared for pregnancy by the action of ovarian hormones. When the developing egg, or conceptus, reaches the uterus an important process of *implantation* occurs. This involves the embedding of a conceptus, at this time called a blastocyst, in the lining of the uterus. The pre-implantation period is critical in all mammals, because it enables both the blastocyst and the uterine lining to mature to the point when implantation is possible. A blastocyst is a small sphere of cells, filled with fluid and containing a small cluster of other cells destined to become an embryo.

Implantation is an essential step in any type of intra-uterine existence. It marks an intimate contact between a conceptus and its mother, necessary for its survival. The process is much more delicately balanced than hitherto realised. It also initiates a series of changes that result in the development of the afterbirth, or placenta. This organ transfers essential substances to, and from, the growing conceptus to the mother. Implantation thus marks the beginning of the parasitic, intra-uterine attachment of the conceptus, an attachment only to be broken, it is hoped, at birth.

The duration of a human pregnancy can be calculated in two ways. The first refers to the *menstrual age* of the newborn child and is calculated from the first day of the last menstrual period. One set of observations gives the duration as $280 \cdot 2 \pm \cdot 3$ days, standard deviation $= 9 \cdot 2$. This is obviously not helpful in a particular instance, but it provides a practical method of estimating the expected date of arrival of the baby in that it represents ten lunar months. To convert these figures so that they refer to calendar months, one goes back three calendar months from the first day of the last menstrual period and then adds on one year and one week. Thus if the first day of the last menstrual period was 1 April, the child can be expected about 8 January the next year. It will be only approximately at that date, for the length of gestation is so variable in man. The chances are about 21–1 against the child being born on 8 January, and for that matter almost the same against it being born on any given day from 31 December to 17 January. But the chances are steadily higher against the child being born each day earlier or later than the inclusive dates given above, and of all children expected on 8 January two-thirds will be born within the range given above.

The age of the child at birth could also be estimated from the time of fertilisation – the *fertilisation age* – but this must rely on the assumption that ovulation occurs about 13–14 days after the first day of the last menstrual period and that fertilisation follows within hours of ovulation; thus the figure of 267 days given for the average length of true gestation is somewhat theoretical.

The reasons for the great variation in the length of the

gestation period are several. Perhaps foremost they involve the hormonal activities of the maternal endocrine organs and those of the placenta, or afterbirth. In women and, as far as is known, to some extent in other mammals, the placenta takes over some of the hormonal functions of the ovary. This enables it to act as a timing device in that its hormonal output can vary in quantity and type of hormone, and should such hormones enter the maternal circulation they may affect the maternal endocrine organs and thus the mechanisms controlling the onset of labour. The output of these hormones almost certainly varies in different individuals, and the response is equally likely to vary. Other factors influencing the length of gestation involve the size of the child, whether it is the first or subsequent, whether it has a twin, and the state of the uterine muscle; certain psychological factors may precipitate labour earlier. Table 2 gives the average gestation periods of a number of mammals in order of length.

Number of young

The term *fecundity* is used to denote the quantitative aspect of the ability to produce young. An animal has high fecundity if it produces many young, it is *fertile* if it is qualitatively able to reproduce. Fertility early in life when also associated with high fecundity tends to persist throughout the reproductive life, and in domestic animals these traits are sought after for the purpose of selection.

The usual number of young born at one time to the human mother is one, and that is also the usual number in all the Primates except in lemurs and marmosets, where two are often born. In one out of every 85 pregnancies in women of Western races twins are born, either like (identical) twins or unlike (fraternal) twins, the latter being born three times as often as identical twins. Triplets are born in one out of every 85×85 pregnancies, and so on to obey what is often called Hellin's law of multiple births. The human uterus is not anatomically satisfactory for more than one foetus, and since twins and triplets are so often born prematurely they tend to suffer a hazardous start in life. There is also a risk (though rare) that

Table 2. *Gestation periods of some mammals. Average length in days, but see notes below.*

Virginia opossum (*Didelphis*)	12·5	Pig (*Sus*)	112
Golden hamster (*Mesocricetus*)	16	Tiger (*Panthera*)	113
Common shrew (*Sorex*)	20	Marmoset (*Callithrix*)	145
Mouse (*Mus*)	20*	Sheep (*Ovis*)	150
Dormouse (*Muscardinus*)	21	Goat (*Capra*)	150
Rat (*Rattus*)	22*	Rhesus monkey (*Macaca*)	163
Mole (*Talpa*)	30	Brown bear (*Ursus*)	210
Rabbit (*Oryctolagus*)	31	Chimpanzee (*Pan*)	226
Weasel (*Mustela*)	35	Hippo (*Hippopotamus*)	240
Hedgehog (*Erinaceus*)	35	Orang-utan (*Pongo*)	260
Grey squirrel (*Sciurus*)	40	Gorilla (*Gorilla*)	260
Hare (*Lepus*)	40	Man (*Homo*)	267†
Ferret (*Mustela*)	42	Porpoise (*Phocaena*)	270
Cat (*Felis*)	63	Common seal (*Phoca*)	275‡
Dog (*Canis*)	63	Jersey cow (*Bos*)	278
Guinea-pig (*Cavia*)	67	Horse (*Equus*)	340
Greater horseshoe bat (*Rhinolophus*)	70	Dolphin (*Tursiops*)	360
Vampire bat (*Desmodus*)	90	Sperm whale (*Physeter*)	365
Leopard (*Panthera*)	90	Giraffe (*Giraffa*)	450
Lion (*Panthera*)	110	African elephant (*Loxodonta*)	660

* This is in females pregnant for the first time; in suckling females gestation is prolonged by delay of implantation.

† Estimated from time of ovulation; from first day of menses the average gestation is 280 days.

‡ Excluding a two-month period of delayed implantation.

in identical twins there may not be complete separation of the pair, leading to some degree of conjoined (Siamese) twinning. Strangely, this seems to occur more frequently in certain countries.

REPRODUCTIVE PATTERNS IN GENERAL

The description given above of the pattern of events in human reproduction raises many academic and practical questions. Is the pattern unique to the human species? What are the patterns in other mammals? What controls the patterns, and can any pattern be altered?

Puberty

There is great variation in the time of onset of puberty in different mammalian species, but in most, though not all, there is some correlation between size or weight and the length of time before the onset of puberty. Thus large whales tend to enter puberty at or after the second year of age; small mammals reach puberty earlier, yet elephants do not reach puberty until nearly ten years old. This indicates remarkable powers of growth in whale foetuses and in young whales, which may well be related to the buoyancy of the medium in which they exist. The state of development of the young of a species at birth may also have an influence on the time of onset of puberty, again bearing in mind the weight and size of each form. A guinea-pig has about the same length of pregnancy (see Table 2) as a cat (63 days), yet a guinea-pig is born in a more developed state than a kitten. The eyelids are open, the permanent teeth erupted and the guinea-piglets altogether more robust. Guinea-pigs are smaller than cats, so it is not surprising that they reach puberty (see Table 3) at 8–10 weeks whereas cats do not do so until 60 weeks old. The guinea-pig is able to carry relatively large foetuses for a relatively long period, because its ovaries produce a hormone, relaxin, which has an effect on the maternal pelvis. The hormone causes resorbtion of the pubic region and allows the birth of a larger foetus at term (the end of pregnancy) than would be possible without its action. There seem to be interesting relationships between size of a female mammal, size of its young at birth, the length of pregnancy, the degree of development of the young at birth and the age at puberty.

Puberty is induced by the anterior lobe of the pituitary, although interrelations with, and feedback from, other endocrine glands may also be involved. Some organs, such as the ovary, start to function in a subdued fashion some years before puberty: ovarian hormone (oestrogen) appears in the urine, with cyclical variation, about two years before the first

Table 3. *Age on reaching puberty.*

Mouse	5–7 weeks	Common pipistrelle	2nd year
Rat	6–9 weeks	Sperm whale	15 months – 2 years
Golden hamster	7–8 weeks	Bush-baby	20 months
Guinea-pig	8–10 weeks	Blue whale	2 years
Rabbit	5–9 months	Pig-tailed macaque	50 months
Dog	6–8 months	Common seal	5–6 years
Sheep	28–35 weeks	Rhesus monkey	3–4½ years
Pig	28–30 weeks	Chacma baboon	4 years
Cow	6–18 months	Brown bear	6 years
Horse	11–12 months	Gibbon	8–10 years
Stoat	1 year	Chimpanzee	8–9 years
Marmoset	14 months	Indian elephant	9–14 years
Cat	15 months	Man	13–15 years

menstrual flow. Puberty is primarily brought about by genetic instructions, but climatic and nutritional factors affect its onset in animals, and in man there are social and possibly economic ones as well. Careful control of diet can prolong the pre-pubertal period in experimental animals, yet the treated animals appear to live no longer than the untreated. Postponement of the onset of puberty in man does not seem to be a practical possibility for increasing his expectation of life. There is, however, an absolutely longer period before the onset of puberty in the larger and higher Primates, irrespective of their size. Man has a longer period of childhood before puberty and a longer period of adolescence between puberty and sexual maturity than all other mammals of his size and all other Primates. In apes, body growth ceases about the age of 12 and bone growth ceases by the age of 14; growth in man continues longer, and some bones continue growing until the age of 22–26.

Reproductive Patterns in other Mammals

A consecutive series of sexual cycles also occurs in the reproductive life of many female mammals, but there are several notable differences from the pattern of events in the human female. There is no external loss of blood, except in certain Primates mentioned below, to mark the end of each cycle and the start of a new one. Overt menstruation occurs in the Old World monkeys (*Catarrhines*) and in the apes (*Pongidae*), but in most New World monkeys (*Platyrrhines*) the uterine changes are not dramatic and bleeding is slight (Table 4).

Table 4. *Reproductive cycles in some Primates.*

Species	Breeding season	Length of cycle*	Length of menstruation
Human	All year	21–34 days	5 days
		Mean 28·32 ± 0·6	Range 2–8 days
		S.D. 5·41	
Spider monkey	All year	24–27 days	3–4 days
Crab-eating macaque	All year	27 days	2–6 days
		Range 24–52 days	Range 2–13 days
Rhesus monkey	All year	28 days	4–6 days
		Mean 27·36 ± 0·17	Range 2–11 days
		S.D. 5·7	
Chacma baboon	All year	41 days	4–9 days
		Range 29–63 days	
Gibbon	Not regular	30 days	2–5 days
		Range 21–43 days	
Chimpanzee	All year	35 days	2–3 days
		Up to 50 days in young	

* S.D. = standard deviation.

The length and number of cycles also vary, and they may not be continuous throughout the year. The period when a female mammal is most receptive to the male during any one cycle is marked by certain organic changes and behavioural manifestations collectively called *oestrus*, or 'heat'; in these mammals there is an *oestrous cycle* (see Table 5 for details).

Any mammal displaying a succession of oestrous cycles in a year is said to be *polyoestrous*, and when the cycles are limited to a certain period of the year, as, for example, from late autumn to early spring in the goat, the animal has a *breeding season*. The interval between two breeding seasons is one of relative quiescence, or *anoestrus*. Some mammals, such as the cat, dog, and ferret, have one long sustained oestrus in the spring, with perhaps another later in the summer or early autumn, and are referred to as being *monoestrous*. The sex cycles in the human female are, in a way, oestrous cycles, except that there are no marked manifestations of heat or, oestrus. Changes do indeed occur in the reproductive organs, but there is no display or any external signs or any appreciable change in behaviour in women, such as are observable in truly polyoestrous mammals. There is no breeding season in a modern human female's reproductive life, yet there is some evidence from the study of primitive peoples that in the past there may have been periods of relatively greater sexual activity at certain times of the year. It has been suggested that May Day, with its ceremony of dancing round a phallic symbol – the maypole – may have been associated with seasonal reproductive activity.

The most important event associated with oestrus is *ovulation*. In the majority of mammals ovulation occurs spontaneously, under hormonal control, either during or just after oestrus. The length of oestrus varies in different polyoestrous animals, from a few hours (at night) in rats and mice, to six days or so in the mare. In monoestrous animals, ovulation is also under hormonal control, but is so set that it needs to be 'triggered off' by the stimulus of mating and is thus known as *induced* ovulation. It may occur at almost any time during a prolonged heat period of up to several months. Should the mating not be a fertile one, there follows a period of false pregnancy (pseudo-pregnancy) during which the reproductive organs enter into a phase of preparation for pregnancy and the animal even behaves as if it were pregnant. Eventually the changes subside and the animal usually enters a period of anoestrus until the next heat. There is some evidence of a condition in the human female which is not unlike pseudo-pregnancy, but it is rare and may well have a strong psychological background.

Animals giving birth to more than one young (a litter) at each labour are called *polytocous*, and *monotocous* animals therefore become parents of only one offspring at each pregnancy. In polytocous mammals the uterus is more often arranged in the form of two long tubes or horns, with foetuses and their

Table 5. *The characteris ics of reproductive cycles in some mammals.*

Species	Type of cycle*	Length of cycle	Length of oestrus	Time of ovulation
Hedgehog	M Spring and summer	—	Not known	Spontaneous
Dog	M Spring and autumn	—	7–9 days	Probably spontaneous
Cat	M Spring and autumn	—	4–10 days	Induced
Rabbit	M Anoestrus October–March	—	Indefinitely in breeding season	Induced
Rat	P All year	4–6 days	20 hours	10 hours after start of heat
Guinea-pig	P All year	16–17 days	6–11 hours	10 hours after start of heat
Pig	P All year	20–22 days	2–3 days	End of heat
Cattle	P All year	18–22 days	24 hours	Just after heat
Sheep	P May have anoestrus in summer	16–17 days	30–40 hours	24 hours after start of heat
Horse	P March–October	20–22 days	6–7 days	End of heat
Rhesus monkey	P All year	27–28 days	Max. receptivity for 2 days	Mid cycle
Chimpanzee	P All year	34–35 days	Max. receptivity for about 4 days	Mid cycle
Man	All year	28·32 ± 0·6 days S.D. 5·41	—	Mid cycle

* M = monoestrous; P = polyoestrous.

placentae arranged like beans in a pod. In monotocous animals the uterus may be of a two-horned or bicornuate type, with the foetus established in one of the two horns, or the uterus may be unicornuate with only one horn, actually compounded of the two fused together. It is strange that besides the Primates, except Tarsius, only bats have a unicornuate uterus.

Large animals are usually monotocous and polytocous animals are usually small, except that most bats have only one young at a time. A large animal has to be born in a relatively advanced stage of development to survive; whales and seals can swim at birth, and fawns can run so fast seven hours after birth that one cannot catch them. Thus the larger the animal, the better the young should be able to look after themselves at birth. Small mammals do not have such problems and can be born relatively undeveloped, and the immediate post-natal period can be spent wrapped in straw or hay or some form of nest. Bats are an obvious exception; their specialised life demands rapidity in learning to fly so as to leave the mother unencumbered. In the human newborn baby we see the prolongation of a 'foetal' type of life after birth, with its concomitant requirements of mother love and parental care for many months. Therefore, in the relatively large human mammal the period of gestation is as long as is anatomically feasible; the young are born in a state far short of being able to fend for themselves, and there is a long period of parental care. This is, of course, associated with the fact that the brain in the adult human is so large that it takes longer to attain full size than in any other mammal. If the human brain were to be more than one-quarter of the adult size at birth, the foetal skull could never escape from the pelvis.

SUMMARY

Reproduction is the process by which animals and plants are replaced. It is also the means by which new forms, differing in some way from their parents, are created and so bring about evolutionary change. Reproduction is relatively simple in the lowest forms and occurs by a splitting into two, binary fission, or by budding. In higher forms, special organs, the gonads, are the site of production of germ cells. Ovaries produce eggs, testes manufacture spermatozoa, so enabling sexual as opposed to asexual reproduction. Spermatozoa unite with eggs at fertilisation to make a zygote in which chemical instructions from the two kinds of germ cells will cause the development of the new individual. The zygote may develop in the sea, inside an egg which may or may not be retained in the mother, or in a special chamber within the mother known as the uterus. Intra-uterine development of the young and the associated evolution of a placenta are characteristics of mammals.

Various factors control reproductive processes, and all have evolved in an ordered and interrelated complexity such that intra-uterine pregnancy is a successful biological phenomenon. Mammals, and man, have to exist as male and female forms which are of a size at birth such that they can leave the mother safely and without damaging her. This means that human babies are not advanced in their development and are, fortunately, quite unable to reproduce in the immediate post-natal period. There must be a period of childhood to allow for growth and further development in many organs, especially the brain. Childhood is long in man and ends at puberty (at an average age of 13·5 years) which is marked by certain anatomical changes, the most important being those in the gonads. Their appearance is controlled by the anterior lobe of the pituitary gland.

Puberty marks the onset of reproductive ability, exhibited in females by ovarian and uterine cycles and in males by the power of emission of spermatozoa. The ovarian cycle is divided into a first (*follicular*) phase, ending at ovulation, and a second (*luteal*) phase, ending in higher Primates and human females at menstruation. The phases are controlled by the pituitary, and through the endocrine activity of the ovaries (follicle and the corpus luteum, which develops in the follicle after ovulation) changes are effected in the uterus and other organs. The ovarian and uterine cycles vary in their details in the different Orders of mammals. In most, the first (follicular) phase is marked by a secretion of powerful oestrogenic hormones which stimulate the female to exhibit 'heat', or oestrus, not overtly present in the human cycle. Oestrous cycles may or may not be limited to a breeding season, not, however, exhibited by man. Ovulation, marking the release of the egg from the ovarian follicle, is usually closely associated with oestrus. The ovarian cycles continue in most mammals until death, interrupted by pregnancy and lactation, but in the human female the cycles stop more or less abruptly at the menopause (average, 49 years of age). Menstruation also ceases at this time and post-menopausal women are, therefore, reproductively sterile. The phenomenon of menstruation involves the breakdown and loss of most of the inner lining, endometrium, of the uterus. There is a repair period after menstruation until the uterine lining has become suitably prepared for pregnancy. The average length of the human menstrual cycle is 28 days, and menstruation lasts from 4–5 days.

Fertilisation of the human egg takes place in the uterine tube shortly after ovulation when spermatozoa reach the egg. The fertilised egg, or zygote, begins to develop at once, and 5–6 days later reaches the cavity of the uterus as a blastocyst. This is a small sphere of cells with an inner cluster marking the undeveloped embryo. The blastocyst becomes attached to the uterine lining at implantation, which marks the true start of the intra-uterine period of gestation. The length of gestation varies in mammals depending on such factors as the number of young, their size and state of development at birth, as well as the type and size of the uterus. Man is usually monotocous (one young born at a time, though twins are born 1 in 85 pregnancies), and the gestation period averages 267 days but is more variable than in any other mammal.

3. Female Reproductive Organs and Their Control

The mammalian reproductive apparatus consists of primary organs, the gonads, responsible for production of sex cells, and of secondary organs forming a genital tract with associated glands. The genital tract is needed for the conveyance of the sex cells to the exterior in males, and for egg transport and embryo development in females.

Lower animals convey their sex cells directly to the exterior into a watery environment such as the sea. This means that their eggs are shed before fertilisation and they are oviparous. Some bony fishes (teleosts) are ovoviviparous: the eggs are fertilised inside the female, in the ovary, where they are retained while the embryos develop and until the young are evacuated from the mother at birth. Higher animals, and some fishes and reptiles, are viviparous with internal fertilisation and retention of the developing young in a chamber in the genital tract which protects them until birth. There is also in association with viviparity an evolution of an organ intermediary between mother and young, the placenta, evolution of mammary glands and a lactation period and also a period of protective care of the young after birth.

Certain other trends are clearly discernible in association with evolution of viviparity. The number of eggs shed from the ovary at any one ovulation is much decreased compared with the profligate release by some fish of thousands or even millions of eggs into the sea at one time. With this marked reduction, to as low as one egg shed at a time in many mammals, there has developed an efficient mechanism for internal fertilisation of the egg and its subsequent transport to the uterus. The maternal endocrine system has become adapted to prolong gestation, and timing mechanisms ensure that birth occurs after an appropriate interval. The foetus, too, displays physiological adaptations, in association with placental functions, which allow its survival in the uterine environment and prepare it for a successful existence after birth. Not unnaturally there are signs of adaptations to viviparity in all mammalian female reproductive organs, and it is remarkable how many are present in forms lower than mammals, although they subserve slightly different functions.

The ovaries of mammals are, in general, paired compact organs and, because of the nature and type of development of mammalian eggs, quite unlike the relatively enormous ovaries of fishes and birds. The ovaries of a large ling (*Molva*) can contain up to 160 million eggs! In mammals the ovaries lie either in the abdomen near the lower pole of the kidney, or in the lumbar region, or, as in women, on the side wall of the true pelvis. They are equal in size, allowing for anatomical changes during activity, except in the duck-bill platypus (the right ovary atrophies) and in some bats (the right enlarges). The ovaries of women are attached on each side to the back layer of a fold of peritoneum stretching out from the sides of the uterus. This fold is the broad ligament, called *alae vespertilionum* by classical anatomists because together the folds resembled bat's wings stretched across the pelvis. Blood vessels pass to the ovary through other folds and mesenteries by which the organ is attached to the broad ligament (Fig. 1). The upper edge of the ligament contains the uterine tube, one on each side of the uterus. Each tube is about four inches long in the human female, is sinuously curved and stretches from an outer frilled end near each ovary to an inner end joining the uterus (Fig. 2). The uterine tube has been referred to for several hundred years as the Fallopian tube in memory of Gabrielo Fallopius (1523–63), Professor of Anatomy at Padua. He was not aware of the function of the tubes as oviducts, but thought of them as tubular chimneys allowing the escape of 'sooty humours' from the uterus. The frilled or fringed ovarian end is referred to as the fimbriated region: it is relatively free,

and its frills almost embrace or caress the surface of the ovary.
The uterine tube conveys the egg after ovulation to the uterus,
fertilisation of the egg takes place in the tube near its
fimbriated end.

The ovary has two principal functions: it stores, matures and
liberates eggs; and it produces hormones that affect profoundly

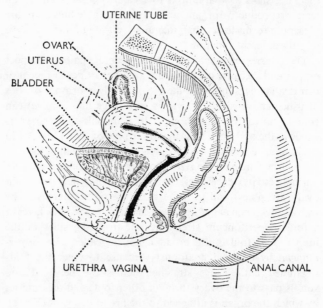

Fig. 1. *Diagram to show the position of the ovary, uterus,
and vagina in a human female.*

the reproductive tract, prepare it for pregnancy and maintain
its usefulness during pregnancy. During reproductive life the
ovary passes successively through a series of stages comprising
an ovarian cycle.

At birth the human ovary contains about a quarter of a
million undeveloped eggs: they are called primary oocytes at
this stage. Formed in man before and just after birth from the
germ cells, they remain in a state of suspended activity for
many years. During this time they are surrounded by an

envelope of epithelial-like cells forming a primordial *follicle*. Processes leading to maturation of follicle and oocyte begin just before puberty. Several primordial follicles start enlarging at the beginning of each oestrous cycle, but in man only one usually reaches maturity and ruptures to release an oocyte. Those that do not reach this final stage become *atretic* and die.

The single-layered maturing follicle soon becomes surrounded by several layers of two kinds of cell. Closely surrounding the oocyte are *membrana granulosa* cells, outside them lie *theca interna* cells. A split develops amidst the former to give rise to a fluid-filled follicular cavity. The theca interna cells become active and secrete ovarian sex hormones called *oestrogens*. In many mammals the theca interna layer becomes so large that it forms a *thecal gland*. It reaches its highest activity just before oestrus. The follicle enlarges rapidly until, within days, it is many times the size of the oocyte. In a human ovary it reaches a diameter of nearly half an inch when mature; it is often known as a Graafian follicle, after a Dutch embryologist, Regnier de Graaf (1641–73).

About the time the follicle develops its cavity, the primary oocyte divides unequally into a large secondary oocyte and one tiny *polar body*. This is the first maturation division, and it (and strictly the next division) results in reduction of the somatic (diploid) chromosome number to half (haploid). The excess chromosomes are extruded in the tiny polar body. Inequality of division preserves cytoplasm within the large secondary oocyte that will be needed during subsequent development. A second maturation division will occur later, resulting in extrusion of another tiny polar body and the formation of the mature ovum. The first polar body may divide again, making three in all; they take no part in development, though it has been reported that they can be fertilised (Fig. 11). The second maturation division occurs in most mammals, and, we suspect, in man, only as a result of the stimulus of fertilisation.

While in the follicle the secondary oocyte becomes enclosed within a transparent egg membrane called a *zona pellucida*. Nearly all types of egg gain protective membranes; three varieties are recognised. One type is secreted from the outer surface of the egg itself (vitelline), another is formed from follicle cells, and the third type is laid down round the egg as a shell while it passes down the oviduct. The mammalian zona pellucida is generally believed to be formed from the oocyte, although many maintain that granulosa cells contribute to it. A protective function has been ascribed to it, but many substances can diffuse through it, probably along minute penetrating processes of granulosa cells.

One or more follicles mature during the first part of each reproductive cycle in human females. Maturation is brought

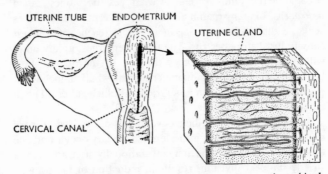

Fig. 2. *Diagram to show the ovary, uterine tube and back of the uterus which has been cut in half. The diagram on the right shows the structure of a small piece of endometrium.*

about by hormonal stimulation from the anterior pituitary. Usually, follicles mature alternately in the two ovaries. The climax of the maturation process is ovulation. The follicle ruptures under hormonal control, and the secondary oocyte is shed surrounded by a zona pellucida and a covering of loose granulosa cells, the *corona radiata*. Cycles can occur without ovulation and are then called anovular.

The shed oocyte is received into the frilled outer end of the uterine tube. In carnivores the ovary is surrounded in a complete ovarian pouch, or bursa, of peritoneum, from which leads the opening of the tube. In some mammals the tube may be attached to the edge of the ovary (ovulation fossa of the mare). The oocyte is transported down the tube by peristaltic movements and by ciliary action that moves mucus in the tube. The oocyte does not always enter the tube; rarely it may be fertilised and develop in the ovary or elsewhere (ectopic or 'out of place' pregnancy).

Mammalian oocytes and eggs are all remarkably similar in appearance. The mature eggs are small, 100–150 μ in overall diameter, with little yolk (miolecithal, as opposed to the heavily yolked megalecithal bird's egg) and the nucleus is near the centre of the egg as opposed to lying near one pole. Once ovulated, their life span seems to be extraordinarily short. Experiments in laboratory animals suggest that viability and thus ability to be fertilised lasts from 12 to 30 hours, after which time degenerative changes appear in the eggs. It is not exactly known how long human ova remain viable, but what evidence there is suggests a period of 24 hours or even less.

After ovulation the spent follicle from which the maturing oocyte was shed undergoes great changes as the result of the action of hormones from the anterior pituitary. The rupture point heals over, and the wall of the old follicle develops into an endocrine organ, the *corpus luteum*, that is able to continue secreting oestrogens and also other important progestational hormones. The principal of these, but not for certain the only one, is *progesterone*, and it acts on the uterine lining and causes its glands to secrete. In a cycle that is not interrupted by pregnancy the corpus luteum persists for only 13–14 days; it then degenerates, and menstruation follows almost immediately.

Although the name of the gland implies that it is yellow (it was first described by de Graaf, who saw it in 1672 in the ovaries of cows in which it is a yellowish, even orange body), it is a pale creamy grey in human ovaries. The gland develops rapidly, and blood vessels grow into it during the first four days. There is often a fluid or blood-filled central cavity at first, but by nine days it is usually solid and up to two centimetres in diameter. A corpus luteum of a cycle soon degenerates and is replaced by a connective tissue scar, it becomes a *corpus albicans* and in about two months hardly a

trace of it is left. In whales the corpora albicantia persist, it is
believed for the length of the whale's life, and thus they can be
counted to give an estimate of the female whale's age.

If the cycle is broken by the supervention of pregnancy, the
corpus luteum persists. The corpus luteum of pregnancy varies
in its fate in different groups of mammals. In some it persists
until term in a functional state. This means that if the ovary

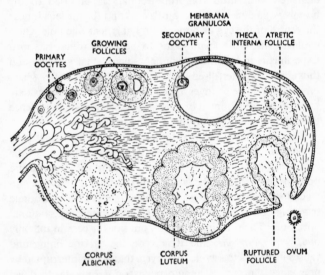

Fig. 3. *Diagram to show the events in the ovary during a
cycle. They occur, of course, in one place in an ovary, but
are shown here in successive stages.*

containing the corpus is removed for any reason during pregnancy, in these forms abortion almost inevitably follows. The
human corpus luteum of pregnancy is functional only for the
first two to three months, after which its activities are taken
over by the placenta, and the gland degenerates. Removal of
the human ovary is, therefore, possible during the later months
of pregnancy without loss of the conceptus. It is, unfortunately,
almost impossible to tell the difference between an active
corpus luteum of the cycle and that of pregnancy. In some

mammals, such as rabbits, cats and dogs, a sterile mating may result in the development of a corpus luteum of pseudo-pregnancy. The gland persists for about half the length of life of the corpus of pregnancy and not unnaturally does not exert a full effect on the uterus.

Should the egg be fertilised, it will reach the uterus when the corpus luteum is fully developed and exerting its maximal influence on the uterine lining. The secretions of the lining are called 'uterine milk', and they provide a nutritious pabulum on which the developing conceptus can live. It will thus be clear that the developing egg must be 'delayed' in its arrival in the uterus until the latter is ready to provide it with nutri-ments. This delay is provided by the length of time (4–10 days in mammals) taken to traverse the uterine tube, and the latter is, therefore, a truly advantageous structure and not simply a conducting pipe for eggs. It is probable that some eggs do arrive in the uterus too early, and die for want of adequate nutriments; and it is quite certain that some do not reach the uterus until too late. In the latter instance the developing egg may have got obstructed in the tube, or perhaps passed down it too slowly and entered too early into its next stage of development before reaching the uterus. Prolonged delay in the uterine tube may result in a tubal ectopic pregnancy, and since the uterine tube is not constructed for containing a rapidly growing embryo or invading placental tissue, the embryo soon dies or else the tube ruptures. Ectopic pregnancy occurs in about 0·25 per cent of all pregnancies in man, but is peculiarly rare in animals. This is certainly partly because some of the obstruction in the tube are caused by disease processes limited to man, and it may also be due to the possibility that higher brain centres, or other factors, may disturb the efficacy of reproductive functions more in the human female.

The human uterus is a pear-shaped organ lying behind the urinary bladder in the pelvis of women who are not pregnant. There is a narrower, lower end, the cervix, which projects into the vagina. The tissue of the cervix is different to that of the uterus; it is composed of connective tissue and is much firmer in consistency. The body of the uterus is composed principally of involuntary muscle, the myometrium, and a relatively thin velvety lining, the endometrium. The functions of the uterine lining are to provide secretions for the developing egg and a layer in which the young conceptus becomes implanted. The function of the thick uterine muscle is to provide protection for the growing embryo and also the expulsive power needed at parturition.

Each convoluted uterine tube is attached to the body of the uterus at its top on the left and right. The lumen of each tube communicates with the uterine cavity by a narrow tubo-uterine junction. The cavity is like an almost triangular slit extending transversely across the uterus. It continues through the narrow cervical canal and opens into the vagina at the external os (mouth) on the cervix. It is this region from which smears are taken to detect early changes in the cells covering the cervix which might lead to cancer. In life the cavity of the uterus and the cervical canal do not really exist, as their linings are virtually in contact except for intervening fluid and, at intervals, the menses.

The vagina is a canal extending from the surface at the cleft between the labia minora, known as the vestibule, to the neck of the uterus. The urethra and bladder lie in front, and the anal canal and rectum behind it. The vagina is directed upwards and backwards and the uterus lies, not in straight continuation, but at an angle directed forwards. This angle varies with the state of fullness of the bladder and, of course, with the degree of dilatation of the uterus in pregnancy. The back wall of the vagina is therefore longer than the front, and there are recesses at the top of the vagina at the back, sides and front of the projecting cervix. These recesses are called the fornices, and their presence enables easier dilatation of the cervix. The walls of the vagina are normally in contact, but they exhibit some transverse folds. The lining is composed of many layers of flattened (squamous) cells. There are no glands in the vagina, and it is lubricated by mucus from glands in the cervix.

Spermatozoa pass up through the cervical canal, through the uterine cavity and into the uterine tube to fertilise the egg in the outer part of the tube. The egg passes down the uterine tube, to die and become absorbed if not fertilised, or if fertilised to undergo a series of preliminary developmental changes before reaching the uterus.

Several words are used synonymously for the uterus, which is Latin for a womb; *uter* means a bag. In Greek the uterus was called *matrix* (relating to mother, thus *metra*, a womb) and *hysteros* (meaning later, because of the later functioning of the organ). Thus in modern terminology we encounter endometrium, for the glandular inner lining, myometrium, for the

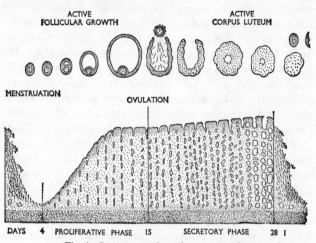

Fig. 4. *Diagram to show the events in the normal ovarian and uterine cycles. The follicles are stimulated to grow by FSH and to secrete and then rupture by LH.*

thick, compact uterine muscle, and hysterectomy, meaning the operation of removal of the uterus. Hysterotomy is the procedure of cutting into the uterus, as is done in more extensive form in a Caesarian section.

The lining of the uterus, the endometrium, reaches maturity at the time of puberty and thereafter passes through a monthly (menstrual) succession of changes in its structure culminating in menstruation. The endometrium is only a few millimetres thick and is composed of numerous glands, blood vessels and supporting stroma, all of which pass through dramatic changes during each cycle. Three layers can be discerned in the endometrium; the two superficial layers are shed at each menstruation, whilst the basal layer remains and from it each new lining is regenerated. There are no nerves in the endometrium, so the pain that often accompanies menstruation is not due to dissolution of the lining tissue. It is difficult to be precise as to what exactly is the cause of the pain; some may be caused by dilatation of the cervix.

After each menstrual period (which lasts two to eight days) there is a stage of repair of the uterine lining, followed by a stage of secretory activity. These stages are successively instigated and brought to fruition by the action of ovarian hormones and possibly by others from the adrenal gland. The hormones are the *oestrogens* – a family of several naturally occurring and powerful hormones that bring about the proliferative phase – and *progestogens*, or progestins, which in their turn and together with the continued action of oestrogens cause the subsequent secretory or luteal phase (Fig. 4). These successive and integrated changes in the uterine lining ensure that it is well developed and active and that its cells possess a chemical composition satisfactory for nutrition of the developing embryo when the latter arrives in the uterus.

Anyone who has kept female pets that are mammals, or who has worked with large female domestic mammals on a farm, will have noticed that there is a blood-stained discharge from their genital openings at certain times. They may well have considered that this loss of blood may be equivalent to menstruation. In fact it is not, but many thought it was until careful studies in reproductive physiology showed that this slight discharge was associated not with the end of one cycle and the beginning of another, but occurred before, or near, the time of ovulation and is often one of the manifestations of oestrus. In the human female there is occasionally a slight discharge associated with ovulation, and even a sensation of pain in the lower abdomen at the same time. These are in no way associated with menstruation, but could be considered as occasionally occurring external manifestations of a subdued or masked oestrus.

Although many of the phenomena repeatedly seen in mammalian reproductive patterns occur in man, it is in that of menstruation that the species differs from all other animals, with the exception of those Primates already mentioned (and perhaps as well the little South African elephant shrew). The phenomenon may play a large part in the culture of primitive races, and is the object of various rites, ceremonies, and taboos, as can be realised from even a superficial reading of J. G. Frazer's monumental *Golden bough* (1890).

It may well be difficult at first to understand the purpose of menstruation; it appears wasteful of tissue and of essential substances such as the iron lost in the blood. The precise cause of the onset of the menstrual flow is not known. It is usually assumed that it results from withdrawal of the action of hormones which built up the endometrium, particularly of progesterone and perhaps indirectly of oestrogen as well. For this and other reasons, it is difficult to find the explanation for menstruation in the anatomy or physiology of the uterus. There seems to be no reason why the uterus should need to lose so much of its tissue, nor at first sight why the endometrium could not simply subside in activity at the end of the secretory phase. The degree of secretory transformation is indeed very considerable by the end of this phase in the human uterus. It may be that the degree of coiling of the vessels, the activity of the glands and the great alteration in the stroma result in so drastic a change in its minute structure that the endometrium becomes 'overripe'. Subsidence in its activity could not occur without death of the tissue. Theoretically it is possible to maintain that in natural uninhibited relationships all the females in a population of animals should be either immature, pregnant, lactating, ovulating or about to ovulate, anoestrous or abnormal. Such a state of affairs almost certainly exists in many large groups of animals in the wild. One could imagine a similar state of affairs in a primitive, short-lived, group of men and women amongst whom intercourse was frequent from adolescence. Menstruation would rarely be seen in such a group and would certainly cause consternation when it did occur.

R.M.—3

THE CONTROL OF GONADAL ACTIVITY

The control of both ovaries and testes is carried out by the pituitary gland. This is a small endocrine gland lying in a recess in the upper part of the base of the skull (the pituitary fossa of the sphenoid bone) and immediately below the middle of the brain. It got its name because it was thought to produce the secretions from a running nose (*pituita* means slime). Only in the past half century have the functions of the pituitary, or hypophysis cerebri (meaning 'growing below the brain'), been elucidated, and there is still much to learn.

The gland weighs little (just over 0·5 g), but its activities have dramatic effects on the body. Without it, life cannot be maintained for long unless substitution therapy is given. There are several distinct parts. There is a median eminence, forming the floor of the central (third) ventricle and contiguous with a small but also important part of the brain, the hypothalamus. This contains several groups of nuclei, some of which at least are involved in reproductive control. There is a thin connecting, hypophysial stalk between the median eminence and the lobular part which is again subdivided. At the back there is a neural or posterior lobe, in direct nervous connection with the hypothalamus by nerve fibres passing down the stalk. In the front is the adeno or anterior lobe. This lobe develops quite separately from the neural lobe as a hollow upgrowth from the roof of the primitive mouth. In embryology it is often called Rathke's pouch after a German worker, M. H. Rathke (1793–1860), who first described its development. The pouch grows up, becomes separated from the mouth and closely attached to the neural lobe. The hollow in the middle of the anterior lobe persists as the hypophysial cleft. The part in front of the cleft is the anterior lobe proper, that part behind which becomes attached to the neural lobe is known as the intermediate lobe. The situation is made a little more difficult in that in some animals the hollow of Rathke's pouch is obliterated, so that there is no hypophysial cleft and no true intermediate lobe. Such an arrangement is found in the human pituitary, and the structure in front of the neural lobe is strictly called the pars distalis. In many animals there is a further continuation of the anterior lobe up round the stalk and even up to the neural eminence. This slender projection is called the pars tuberalis.

The pituitary secretes nine distinct hormones and it can be seen from Fig. 6 that two are produced from the posterior lobe, one from the intermediate lobe and six from the anterior lobe.

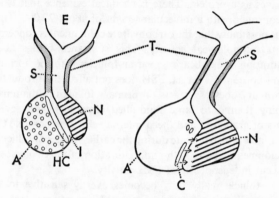

Fig. 5. *Diagrams to show (left) the conventional arrangement of a mammalian and (right) a human pituitary.* E: *median eminence.* S: *hypophysial (neural) stalk.* T: *pars tuberalis.* N: *neural or posterior lobe.* I: *pars intermedia.* HC: *hypophysial cleft.* A: *anterior lobe, pars distalis in man.* C: *marks the occasional colloid cysts marking what might be the remnants of the hypophysial cleft in man. The arrows show the invasion of the neural lobe by cells from the pars distalis.*

Although it could be argued that all of these nine hormones are concerned in one way or another in making reproduction possible, in the strictest sense only certain of them are of primary importance in that they directly affect reproductive organs.

The anterior lobe contains several types of cells. Some stain characteristically with certain dyes and are designated by their tinctorial reactions, and by their types of cytoplasmic granules, as chromophils; others lack certain characteristics and are the chromophobe cells. There is evidence that certain hormones are produced by particular chromophils (known as a, β cells, etc.). A cell type increases in number when a particular hormone is secreted in excess. In some mammals special cells appear during pregnancy and lactation.

The anterior lobe produces three trophic substances (probably from β cells) which stimulate and control gonadal activity. There is follicle-stimulating hormone (FSH) which behaves as a protein. There is chemical evidence that it is a glycoprotein with a molecular weight of about 70 000. There is the possibility that the carbohydrate component is concerned in its physiological activity. This substance brings about maturation of egg-bearing ovarian follicles with the hormone blood-borne to its target. FSH does not affect the production of primary oocytes; it acts on primary follicles in mammals, turning them into large fluid-filled structures with several layers of granulosa and theca interna cells. It is possible that other factors also operate during the earliest stages of follicular development. The growth and maturation of the ovum seems to occur independently of FSH activity.

The follicle really only becomes overtly sensitive to the hormone when it has acquired several layers of granulosa and theca cells about it. Administration of FSH to immature mammals brings on maturation of the ovary and leads to marked sexual precocity. The amount of the hormone secreted fluctuates rhythmically during reproductive cycles; and although it does not in its purest form itself stimulate ovarian hormonal secretion, without its initial preparatory action on the follicles such secretion does not occur.

A second glycoprotein that has been isolated from the anterior lobe is a luteinising hormore (LH). When administered to mature experimental animals it brings about ovulation of some ripe follicles, haemorrhage into others, and the formation of corpus luteum-like tissue (thus luteinising) in medium and large follicles without their rupture. LH acts on ovaries already stimulated by FSH and has a particular effect on the theca interna cells of developing follicles, making them swell and actively secrete oestrogen. A sudden elevation in LH production causes ovulation, and the continued action of LH converts the ruptured follicle into a corpus luteum. There are fluctuations in the amount of LH produced throughout each oestrous or menstrual cycle, and there is also some evidence that this is one of the dominant factors in sexual periodicity. It is also known that there is an interaction between LH and FSH

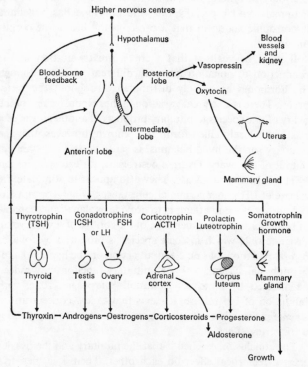

Fig. 6. *Diagram to indicate the hormones produced by the pituitary and their biological activities (modified from C.H.Li).*

that leads to greater stimulation than might be expected from a simple summation of the effects of the two hormones. This interaction is made use of to procure intense artificial ovarian stimulation experimentally, and it probably also occurs at ovulation in normal cycles.

The third trophic hormone isolated from the anterior lobe

is called luteotrophin (LTH), which is probably, but not absolutely certainly, identical with lactogenic hormone and prolactin. Its activities have not been precisely proved or explained, and there is evidence that in some mammals LTH acts only in synergism with a luteinising hormone. In some mammals, and especially rats, the hormone has a definite luteotrophic action in that it prolongs the life of the corpus luteum.

It must be realised that there is great variation in the production of gonadotrophins in different species, although the hormones are widely distributed throughout the vertebrates. There is also great variation in the responsiveness of the ovary of one species to pituitary products of another, even of a relatively closely allied form. It is known, however, that the pituitary, or the hypothalamus, is sensitive to the hormonal output of the ovary. Ovarian oestrogens, for example, inhibit FSH synthesis and release. They also appear to stimulate the release of LH under certain circumstances, which concept has been a basic one in attempts to provide explanations, capable of being verified experimentally, of the control of sexual cycles. Progesterone, which in some species is produced by follicles *before* ovulation, has no clear-cut action on the hypothalamo-hypophysial system. Although pituitary function may indeed be affected by large concentrations of progesterone, with inhibition of LH release, it seems that at low concentrations a synergistic action with oestrogens must occur to bring about a triggering effect.

The conclusion reached is that the pituitary and the ovaries exert a reciprocal effect on each other. There is a 'negative feedback' action of the target organ, through its own secretions, on the pituitary. FSH matures the follicle, FSH and LH cause it to produce oestrogens which inhibit FSH and stimulate LH. Ovulation, luteinisation and progesterone production result, with LTH possibly maintaining the corpus luteum until the cycle starts again or pregnancy results.

The relations and interplay of pituitary and gonads take place within an internal environment in the female animal, which is also influenced by other endocrine glands, such as the adrenal and the thyroid, by other products of the pituitary, such as oxytocin production stimulating LTH release, and by environmental factors such as food, climate, light and the season. There is, as it were, a permissiveness about gonadal control by the pituitary, conditioned at various ages and subject to many subtle influences acting through pathways not yet completely clear.

One of these pathways which has been investigated is that connecting the hypothalamus to the hypophysis, and neither is it nor its mechanisms understood thoroughly. The hypothalamus consists of several groups of nerve cells (hypothalamic nuclei), and there are both nervous and vascular connections with the pituitary. The former comprise numerous fibres of hypothalamic origin which sweep down the pituitary stalk towards the neural lobe. The vascular connection involves a portal system of vessels, that is to say, an independent system of small venous channels originating in the hypothalamic region and passing down to supply the anterior lobe below. One concept proposes that there exists a relay chain of control from the hypothalamus to the pituitary which involves both nervous and vascular links. Appropriate stimuli arrive at the hypothalamus, and their implications are sorted out in relation to the existing levels of pituitary hormones. Nerve impulses travel from the hypothalamic nuclei to the median eminence and there liberate chemotransmitters. These enter the portal plexus and are thus distributed to the anterior lobe activating the secretory cells there.

Oestrogens and progesterone

The physiological oestrogen produced by the ovary is probably 17-beta oestradiol. There are two oestrogens which are thought to be metabolites of oestradiol. They are oestrone and oestriol, and they occur in blood and urine. Oestrogens appear to be formed from cholesterol. The ovaries produce only a few microgrammes each day, and after liberation into the blood stream oestrogen is inactivated rapidly, mainly by the liver. Oestrogen may circulate in the form of 'pro'-oestrogen which is inactive until converted into an active form by certain tissues. The naturally produced oestrogens cannot be taken by mouth and have to be injected to procure any effect. Synthetic oestrogens (ethinyloestradiol, hexoestrol and diethylstilboestrol and many others) have been developed to overcome this difficulty. Most oestrogens in suitable amounts exert an effect on the anterior pituitary and inhibit ovulation; there is no certain correlation of their hormonal and ovulation-inhibiting potency.

Oestrogens exert a physiological action far wider than that directly influencing the reproductive organs. They can affect protein and carbohydrate metabolism; they also play a part in the calcium balance. The relatively more rapid growth of girls at puberty and the earlier union of the epiphyses of the bones in young women than in young men may be attributable to ovarian oestrogens. Oestrogens also affect retention of water and salts in the body. This can again be reflected by changes in the body at puberty and by the sometimes detectable swelling of tissues during the oestrogenic phase of the cycle. Oestrogens affect other endocrine glands as well as the pituitary, the adrenal and the thyroid being especially responsive. They also influence the production of pituitary hormones other than the gonadotrophins. The administration of oestrogen can reduce the production of growth hormone.

Exactly how oestrogens, and other steroid hormones, act is still being discussed. They may alter rates of synthesis of enzymes, activate enzymes, act as cofactors in an enzyme system, or alter cellular permeability. One general theory has suggested that they regulate the genetic programming in cells to cause enzyme formation and thus alteration in the metabolism of cells. Naturally there has been much searching for anti-oestrogens, especially for fertility control. Besides the large series of synthetic steroids having such an effect, anti-oestrogens have been obtained from alfalfa leaves and pine needles. Research continues to discover an effective anti-oestrogen with no side effects and suitable for use by human beings.

A true progestin is defined as a substance which acts on the endometrium to induce proliferation, on the uterine muscle to

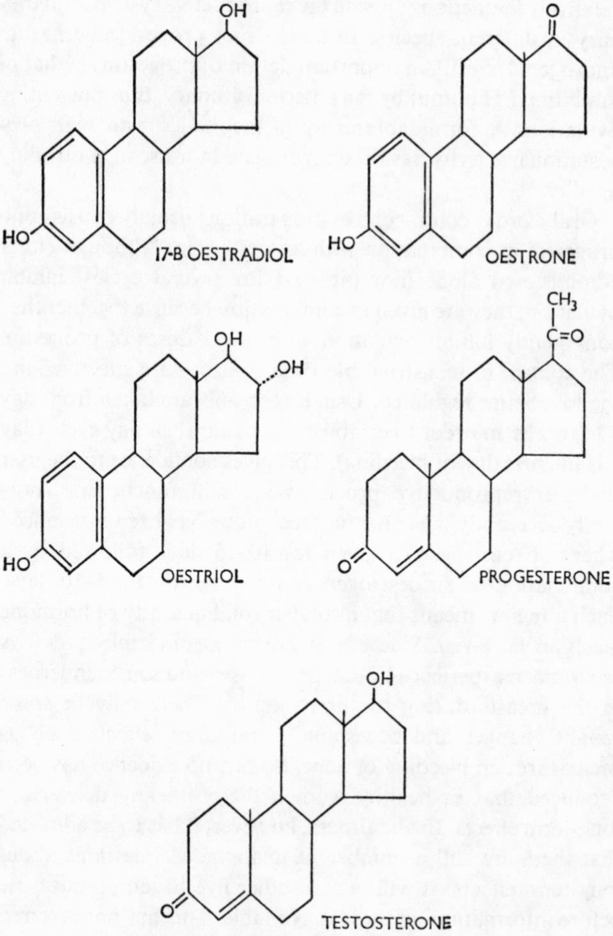

Fig. 7. *Formulae of steroid hormones produced by the ovary (oestrogens, progestins) and testis (androgens).*

produce activity characteristic of pregnancy, and which also maintains pregnancy in a uterus containing conceptuses after the ovaries have been removed. Oestrogens may be necessary to enhance the effect of insufficient progestin. Progestins have relatively few actions outside the reproductive system, and they vary in different species. In large doses progesterone has an androgenic effect. An important action of progestins is that of inhibiting LH output by the anterior pituitary, thus preventing ovulation. A formidable array of synthetics with high pro-gestational activity has been investigated for use in controlling fertility.

Oral steroid contraceptive preparations usually consist of a progestin in combination with an oestrogen. Although either administered alone may (at least for several cycles) inhibit ovulation, they are given in combination because together they consistently inhibit ovulation with lower doses of progestin. The control of menstrual bleeding is also more effective, and the loss better regulated. Usually the pills are taken from day 5 to day 25 in order to establish an artificial 28-day cycle (day 1 is the first day of bleeding). This gives confidence to the user that her reproductive processes are still functioning 'nor-mally'. Recently a 'sequential' technique has been introduced where an oestrogen is given for 10–15 days followed by a combination of an oestrogen and a progestin for 5–10 days. Such a regime means that a smaller total quantity of hormone needs to be given. Side effects are minor and infrequent. A few women experience headaches, nausea and some tenderness in the breasts during the first months. There may be some weight changes and occasionally 'nuisance' effects such as break-through bleeding or acne. So far, no evidence has been produced that in healthy women there are any deleterious long-term effects. In all fairness, however, it has to be admitted that there are still a number of unanswered questions about long-term effects. It will take another five to ten years of use before information on these is available. This has not deterred the ten million and more women who now use oral con-traceptives, and the present evidence suggests that they should not be deterred. The chances of becoming pregnant on ceasing to take oral contraceptives are not affected, in fact they may be increased during the first six months of return to natural cycles. Practical advice on oral contraception is best obtained through family planning clinics or from physicians.

SUMMARY

Viviparity, giving birth to live young, demands the possession by a female mammal of primary and secondary reproductive organs within the body. The primary organs are the ovaries. They are responsible for storing, maturing and shedding the female germ cells, and also for producing ovarian sex hormones, oestrogens and progestogens, which have actions on the secondary reproductive organs. The germ cells, the oocytes, are present in large numbers in the ovaries at birth. They mature within ovarian follicles, the walls of which secrete oestrogens. One, occasionally two, or even three follicles mature during each ovarian cycle and rupture at ovulation to release an oocyte. The ruptured follicle subsequently develops into a corpus luteum that persists for 13–14 days and is a source of progesterone as well as oestrogen. The activity of the corpus luteum is responsible for the more stable part of the ovarian cycle in that menstruation follows almost always 13–14 days after ovulation, whereas the early part of the cycle is more variable in length. Unless pregnancy supervenes, the corpus luteum involutes rapidly into a connective tissue scar, a corpus albicans. If pregnancy does occur, the corpus luteum persists, preventing menstruation and maintaining the functional activity of the endometrium.

The secondary female reproductive organs include the uterine tube, uterus and vagina. The convoluted tube stretches on each side from a fimbriated ovarian end to the tubo-uterine junction. It conveys the oocytes; fertilisation of the oocyte occurs in its outer third; it 'delays' the arrival of the zygote in the uterus until the endometrium has been adequately prepared by ovarian hormones. The muscular uterus has an endometrial lining which is made to proliferate by oestrogens. The uterine glands are stimulated to secrete by progesterone after the endometrium has been primed by oestrogens. When the activity of the corpus luteum declines, the greater part of the endometrium degenerates and is shed at menstruation. There is thus a uterine cycle of endometrial proliferation, secretion, destruction and repair which is controlled by the follicular and luteal phases of the ovary. The purpose of the changes is to build the endometrium up into a condition when it can receive the developing conceptus. This state appears only to be tenable with full activity of the corpus luteum. On its subsidence the endometrium is so over-developed that it cannot involute and so degenerates at menstruation.

The control of the ovarian cycle is mediated through the anterior lobe of the pituitary. This small endocrine gland (weighing about 0·5 g) lies below the brain, to which it is attached by a short stalk connected to its posterior lobe. It produces several important hormones, of which the gonadotrophins are primarily concerned with stimulating ovarian activity. A follicle-stimulating hormone (a glycoprotein) acts on primary follicles and causes them to mature. A luteinising hormone acts on follicles already stimulated by FSH, causing them to secrete oestrogens and eventually to rupture with the liberation of the oocyte and the development of a corpus luteum from the follicle wall. A third trophic hormone, luteotrophin, prolongs the life of the corpus luteum in some mammals. There is a feedback effect on pituitary activity by the ovarian hormones and, with the resulting interplay and enhancement by the hormones concerned, an ovarian cycle is established. The events in the ovaries exert an effect on the lining of the uterus and also on other organs, such as the mammary glands. Additional factors can affect the reproductive cycle. They include other endocrine organs, environmental factors such as food, climate, light and the season of the year. There is experimental evidence that the hypothalamic nuclei play a part in controlling anterior lobe activity. Vascular connections link the two regions, and it is suggested that one or more chemotransmitter substances pass to and activate the cells of the anterior lobe. The hypothalamic nuclei may be sensitive to blood-borne hormones from both ovaries and pituitary and also to other stimuli, and will thus be stimulated to send appropriate chemical instructions to the pituitary.

4. Male Reproductive Organs

In man the testes are suspended in a thin-walled sac of skin, the scrotum (Latin for a bag), into which they descend just before or at birth. Until puberty is reached the male gonads, the testes, are immature and neither produce germ cells (spermatozoa) nor act as effective endocrine organs.

The testes of children must, therefore, mature in two ways to become functional reproductive organs. Part of the organ is responsible for producing spermatozoa and another part for secreting the male sex hormones, *androgens*. These are needed not only to stimulate sperm production (spermatogenesis) but also to bring about the development of the characteristic male body and build. As in the female, there is a gradual development of gonadal activity, which starts before puberty and culminates in the ability to ejaculate semen. At the same time the testis is stimulated by the pituitary to secrete increasing quantities of androgens. It has been known since early times that castration of a boy just before puberty prevents the development of the secondary sex characteristics of the adult male. The castrated individual, or eunuch, retains an adolescent build and does not exhibit the distribution of hair, changes in the voice due to enlargement of the larynx, and certain other structural features of the adult male. The testicular hormones are responsible for stimulating the development of the male secondary sex characters, and the general pattern of mammalian reproduction demands that the male should have certain anatomical attributes. The gonads must be able to produce fertile and motile germ cells, contained in a fluid medium and capable of being deposited within the female genital tract by a copulatory organ, or penis.

Man is an example of a 'constant breeder'. From puberty

until late in life the human testis produces spermatozoa continuously; there is, however, some evidence of a slight increase in activity at certain times of the year. Amongst other male mammals, rats, guinea-pigs, and most Primates are capable of fertile sperm production all the year round, but many, such as hedgehogs, squirrels, many Carnivora and Artiodactyla have either a distinct breeding season (or rut), or show seasonal fluctuations in the quantity of semen produced. Reproductive activity subsides gradually in ageing human males; there is no distinct 'change of life' comparable to the menopause in the female. The glands that produce the seminal fluid, the *prostate* and *seminal vesicles*, are known as the accessory male reproductive organs. Spermatozoa need to be mixed with seminal fluid before they become motile. Until they are expelled to the exterior, they are stored in a long narrow coiled tube, which forms a compact structure, the *epididymis*, lying along one border of each testis. The word means 'that which is on the twins', the twins being the testes; in the plural it is spelt epididymides.

Copulatory organs, capable of being inserted into the female vagina, are present in many reptiles, some birds and all mammals. They consist of elongated masses of sponge-like erectile tissue that can become engorged with blood. Some mammals, such as bats, seals and dogs, possess a stiffening bone, the os penis, or *baculum*, as well. The age of these animals may be gauged by the size of the baculum. The penis is traversed by the tubular *urethra* that carries out the double function of conveying urine and, at other times, the seminal fluid with the spermatozoa.

The two functional parts of a mammalian testis consist of a tubular apparatus amongst which lie abundant groups of secretory cells producing male sex hormones. These groups of cells constitute the interstitial tissue of the testis. This tissue was first accurately described by Franz Leydig as long ago as 1857 and is often called after him. He was an enthusiastic comparative histologist, but he did not realise the endocrine function of Leydig tissue. The numerous tubules (800) are coiled and closely packed within lobules. They are called seminiferous tubules, and several lie together within the septa separating the lobules. Their arrangement results in a great increase of tubular surface within a solid ovoid organ. The testis is surrounded by a tough, virtually inelastic envelope of

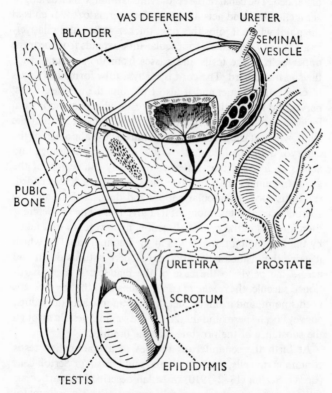

Fig. 8. *Diagram of the reproductive organs in a human male.*

connective tissue, the tunica albuginea, which keeps the inside of the organ under pressure. Fluid excreted into the tubules sweeps sperm made inside them along the tubules to straight collecting ducts and thence to a network of more small tubules (rete testis). From this network some 20 small efferent ducts convey the sperm into the main canal of the epididymis. An important function of the duct system is to absorb the fluid excreted into the seminiferous tubules, otherwise the epididymis would accumulate large quantities and become distended. The canal of the epididymis is about 6 metres long, it is much coiled and acts as a store for spermatozoa. It ends at the point where it joins the long, thicker walled and relatively straight *vas deferens*. This muscular tube extends from near the upper pole of the testis and passes from the scrotum inside the spermatic cord. The cord is really a tube formed of layers of connective tissue, and it also transmits the arteries, veins, nerves and lymphatic vessels of the testis which pass to it from inside the abdomen. In reality, what has happened is that the testis, which started its development inside the abdomen of the embryo, descended down the back wall of the abdomen and across to a canal that traverses obliquely the region of the groin (inguinal canal) to enter the scrotum. As it descended, the testis took its vascular supply and ducts with it, and the spermatic cord is testimony of its movement. The vas deferens therefore lies in the reverse direction, as it were, to that taken by the descending testis. It lies in the inguinal canal by which it gains the inside of the abdomen. There it turns down and across the pelvis to the base of the bladder. A many-partitioned saccule, the *seminal vesicle*, drains into the vas deferens at this point, and a single duct, the common ejaculatory duct, conveys both sperm and fluid from the seminal vesicle through the substance of the prostate into the urethra.

At birth the seminiferous tubules of a mammalian testis contain germ cells, spermatogonia, and 'nurse' or Sertoli cells (Enrico Sertoli (1842–1910) of Milan described 'di particolari cellula ramificate dei canalicoli seminiferi'). The spermatogonia begin to divide just before puberty into primary spermatocytes, and these in their turn divide into secondary ones. After puberty the process continues further by the formation of spermatids from secondary spermatocytes and by their maturation into spermatozoa. In contra-distinction to affairs in the ovary, gametes are actively formed in the testis throughout reproductive life. One primary oocyte gives rise only to one

EPIDIDYMIS VAS DEFERENS

COLLECTING TUBULES

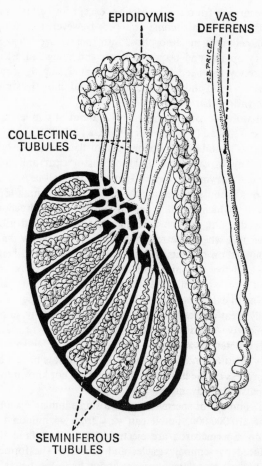

SEMINIFEROUS TUBULES

Fig. 9. *Diagram of a testis showing the seminiferous tubules, and the epididymis and vas deferens.*

ovum, but one primary spermatocyte can form four spermatozoa. Reduction in chromosome number occurs at the maturation divisions, and spermatozoa will contain *either* an X *or* a Y chromosome as well as autosomes.

Human spermatids are oval cells that do not differ much

in appearance from spermatocytes, except for certain nuclear details. They soon become attached, however, to the tips of Sertoli cells and metamorphose into spermatozoa. They lose their cytoplasm and become narrow and elongated. Electron microscopy has been of assistance in analysing the complex changes in form and in providing information on the structure of the various parts.

A mature human spermatozoon is about 54 μ in length and possesses a head (5·0 $\times$ 3·5 $\times$ 2·5 μ), a neck (0·5 μ), a middle piece (4·0 μ) and a long slender tail, or flagellum. The head is covered by two caps (acrosomal and post-nuclear) and contains a nucleus carrying the chromosomes. The small neck is in the form of a granule connecting the head to a spiral middle piece, or body, which has an axial filament with mitochondria arranged round it. The axial filament is continued into the slender tail, where electron microscope studies show that it is surrounded on the outside by a spiral sheath. The tail ends in a leash of five terminal fibrils. Electron microscopy has revealed a more complicated structure in spermatozoa than was believed to exist from other studies. Mammalian spermatozoa display great similarity in their general morphology, although comparative electron microscopic studies may reveal more subtle differences. All have a head, middle piece, and a tail that serves as an organ of locomotion. Those of rats have a hook-like process on the head, but this feature is not known to have any particular advantage.

The number of spermatozoa in a single human ejaculate is about 210 000 000, but it can vary from 45 million to 730 million. Spermatozoa are ejected in a seminal fluid that is produced by seminal vesicles and prostate. The former are two hollow sacculated organs, about five centimetres long, found at the base of the bladder and leading by short ducts to join the tubes conveying spermatozoa from the testes (vasa deferentia) just before they enter the urethra. Seminal vesicles are not present in Marsupialia or Carnivora, but are particularly large in some Insectivora, such as hedgehogs. The prostate is a glandular organ, three to four centimetres in diameter, placed at the neck of the bladder around the first part of the

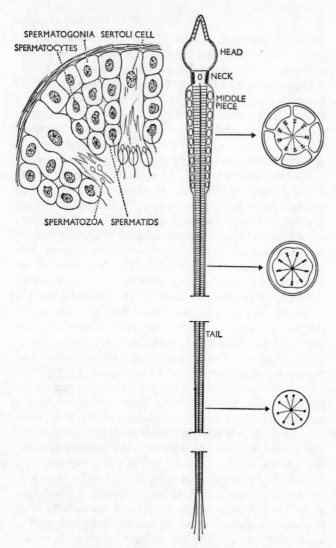

SPERMATOGONIA SERTOLI CELL
SPERMATOCYTES

HEAD

NECK

MIDDLE
PIECE

SPERMATOZOA SPERMATIDS

TAIL

Fig. 10. *Diagrams to show the wall of a seminiferous tubule and right) a mature spermatozoon as seen by electron microscopy. Cross-sections of the middle piece and tail on the right.*

59

urethra. Its secretion is squeezed out by its smooth muscle through a series of ducts into the urethra, where it is mixed with secretions of seminal vesicles and with spermatozoa.

The prostate may be of a 'disseminated' type, such as is found in sheep and goats, in which it is represented by a diffuse aggregation of urethral glands. In most other mammals it consists of two or three lobes, but there may be disseminate tissue as well; only in man and dog does the prostate virtually surround the urethra. The gland is absent in Monotremata.

The total volume of fluid in one human ejaculate averages 4·0 ml (range 1·0 – 11·0 ml). In those mammals without seminal vesicles the volume of seminal fluid is small, as in dogs; in a boar, with large seminal vesicles, it is abundant. Functions of the fluid are to permit survival of sperm in a watery medium after they have left the male, to enable them to be motile, and to provide them with nutriment. The fluid contains all substances necessary for these purposes and also, somewhat strangely, the carbohydrate fructose. This is usually found in plants, and its presence in the seminal fluid is as curious as it is in the amniotic fluid of certain mammals. Fructose does not seem to be necessary for spermatozoa to live on, but it may give seminal fluid a composition and osmotic pressure that provide sperm with an optimal environment.

Spermatozoa undergo final maturation in the coiled epididymis and the vas deferens (sperm withdrawn from a testis are infertile). If not ejaculated they soon lose their potency, degenerate, and may be absorbed. They become motile when mixed with seminal fluid and can remain motile for 48–72 hours within the female genital tract; they may not, however, be fertile for all this time. The tail performs undulatory movements and causes the head to progress forwards with an oscillatory motion along a spiral path. Spermatozoa swim by preference against mucous currents and against gravitational pull. They take about an hour to reach the end of the human uterine tube from the cervix, but not many have arrived there by the end of this time. Performance is affected by conditions prevailing in the female genital tract and by the degree of its muscular contractions. Activity of spermatozoa is their basic, defining characteristic; abnormal spermatozoa without tails are incapable of effecting fertilisation of a passive ovum.

Metabolism of spermatozoa is of great importance in connection with artificial insemination. Their longevity can be increased by cooling; sperm of prize bulls may be collected and transported by air to distant countries. Sperm could be preserved by cooling until after death of the donor, when by artificial insemination offspring would be obtained that were children of a long-dead father (telegenesis). This may be of practical use in improving farm or racehorse stock, but with man it is only for imaginary 'brave new worlds'.

Spermatozoa are responsible for fertilisation of an ovum. This marks the true start of the development of a new individual from a zygote. A fertilised ovum has the appropriate number of chromosomes for its species and also has its future sex determined. Adult cells in a particular mammal each possess a number of chromosomes (the diploid number), the number depending on the species (rabbit, 22; rat, 42; sheep, 54; cow, 60). Whatever the number of chromosomes in the nuclei of cells of adult animals, there is half that number (the haploid number) in the nucleus of each egg and also in the head of each spermatozoon. When a spermatozoon fertilises an egg the original number of chromosomes is restored in the nucleus of the zygote. This reduction by half in gametes of the adult number of chromosomes is essential, otherwise when fertilisation occurred there would be a steadily increased number of chromosomes and eventually there would be nothing in the world but chromosomes. It is known that the number of chromosomes in adult human cells is 46, and that one male chromosome differs considerably from its counterpart in the female. The female chromosome number is composed of 44 chromosomes, or *autosomes*, arranged in pairs, and two identical sex or X-chromosomes. One of the sex chromosomes in male cells is similar to the X-chromosome of the female, but the second is very much curtailed and is called the Y-chromosome. The halving which takes place during formation of germ cells in the female will therefore result in an egg always having 22 + X chromosomes. In the male, however, there will be two kinds of spermatozoon – one possessing 22 + X and the other 22 + Y chromosomes. The genetic sex of an individual will therefore depend on whether an egg (containing 22 + X chromosomes) is fertilised by a spermatozoon containing 22 + X or one containing 22 + Y chromosomes. Should the former occur, then the reconstituted number will be 44 + 2X and the embryo will develop into a female adult. On the other hand, if the spermatozoon fertilising the ovum contains 22 + Y chromosomes, then cells of that embryo will have 44 + X + Y chromosomes and will become male (see Fig. 11).

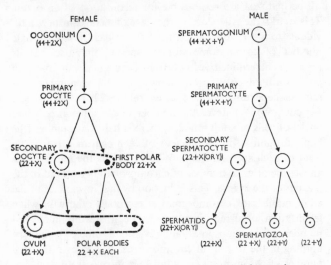

Fig. 11. *Plans to show the chromosomal changes that occur during the formation of a mature ovum and of spermatozoa.*

It can thus be seen that there is a chromosome difference between the two sexes. This genetic difference ensures that in most mammals the form, shape and anatomy of males differ considerably from those of females. Technically, therefore, we could say that the human female is true-breeding, and that because of the Y chromosome the male is a hybrid. The two sexes are obviously well adapted as regards their ability to reproduce, but the genetic difference between the sexes expresses itself in various ways. Not only does it result in sexual dimorphism, but also in certain sexual attributes. The human male, for example, is more variable in his anatomy and is particularly variable in his ability to survive, simply because of his maleness. The death rate of male foetuses is higher than that of female, and even after birth the male is more liable to die. It has been remarked often during the last 2000 years that it is males that are more often born deformed or defective. From a genetic and from other points of view, it is absurd to talk of the two sexes being equal.

Unfortunately, successful fertilisation and the concomitant determination of the sex of the future embryo are not necessarily a guarantee of the birth of a perfect individual, even if the hazards mentioned above are avoided. Not every embryo's cells collect the right number of chromosomes: there may be additional autosomes or sex chromosomes. This may well be due to a faulty division when the germ cells are formed, so that perhaps a spermatozoon carries $22 + X + Y$ chromosomes, or an ovum is formed with $23 + X$. There is also the possibility that some factor, such as age of the ovum, or the effect of chemical or physical influences, interferes with the correct genetic action of the chromosomes. Individuals may be born without the fullest development of the attributes of one sex or the other: the sexual organs may not have developed properly, they may not function as they should; or there may not appear the appropriate type of behaviour, sex drive or libido expected of males or females. Although sex is determined at fertilisation, there must follow the co-ordinated embryological differentiation of male or female organs, the maturation of the proper physiological and endocrinological activities of each organ, and, finally, the integrated psychological manifestations of each sex.

Is there such a concept as a male or a female type of brain? Do female sex hormones cause male brains to have feminine thoughts? Have homosexual individuals acquired the wrong genetic make-up? These questions are exceedingly difficult to answer, partly because of our lack of knowledge and partly because of the difficulties in making the right kind of analysis and experiment. It has been stated that no sex hormone can affect any *adult* central nervous system to behave differently. Sex hormones only stimulate target organs sensitive to them. Recent research has suggested that the situation may be the other way round, in that *female* brains, hypothalamus and pututitary only drive *female* sex organs properly. Male sex organs, at least in some species, appear to be more independent of central control, and to influence the male hypothalamus less by negative feedback of their secretions once they have become functional. Perhaps this indicates why the male lacks the periodicity in his reproductive potential that is so characteristic of females.

It is possible to determine the presence of what is probably the 2X chromosomes in certain female cells by a fairly simple microscopical investigation. In mammals (especially in Carnivora and Primates) the sex chromatin in the cell nuclei of females aggregates in well-prepared material, sometimes near the nuclear membrane, and is more conspicuous than in males. The sex chromatin is better discerned in some types of cell (those with vesicular nuclei) than in others with dense nuclei. It is clearly seen in many neurons, and also in cells obtained by skin biopsy, by buccal and vaginal smears, and by urine sediments. The neutrophil leucocytes, obtained from peripheral blood and examined in impeccably prepared smears, also exhibit sexual dimorphism in a proportion of their number. There is a striking drumstick-like nuclear appendage with a head $1\cdot4\,\mu$ to $1\cdot6\,\mu$ in diameter attached by a fine thread to the nucleus. The methods can provide information about the genetic sex of individuals with abnormal reproductive organs (p. 77).

TESTICULAR ACTIVITY. DIFFERENTIATION
OF REPRODUCTIVE ORGANS

The male gonads are under the control of the anterior lobe of the pituitary, just as are the ovaries, but without the same periodicity. FSH is produced from male anterior lobes, and it causes growth of the testis, increase in size of seminiferous

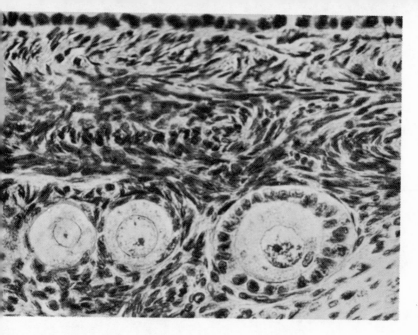

Plate I. Above: *Photograph of a section through the cortex of the ovary of a monkey showing oocytes within young follicles.* Below: *Cross-section of a growing follicle in a mammalian ovary showing the membrana granulosa and the theca interna developing about the oocyte.*

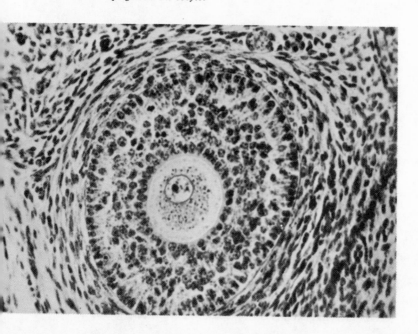

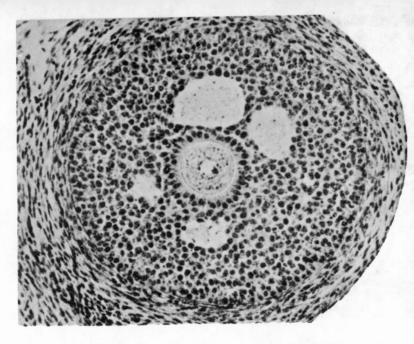

Plate II. Above: *Section of a growing follicle from a mammalian ovary. The follicular cavity is forming and the zona pelucida is being laid down about the oocyte.* Below: *A mature follicle in a mammalian ovary just before ovulation.*

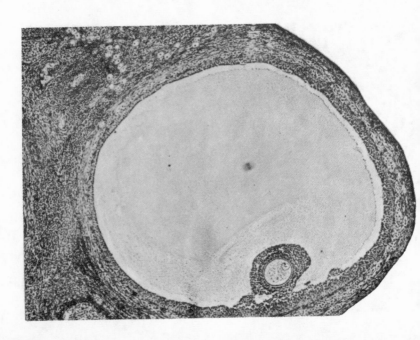

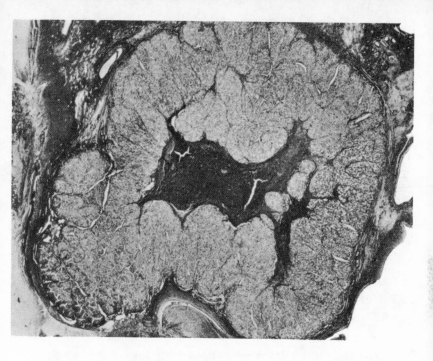

Plate III. Above: *Section of a well-developed human corpus luteum.* Below: *Section of a human corpus luteum a few days old.*

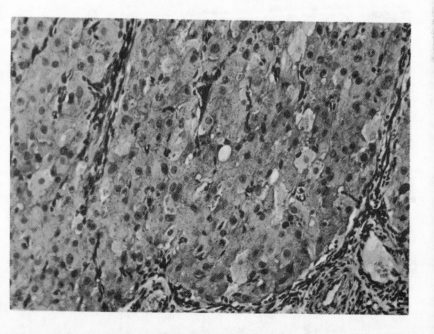

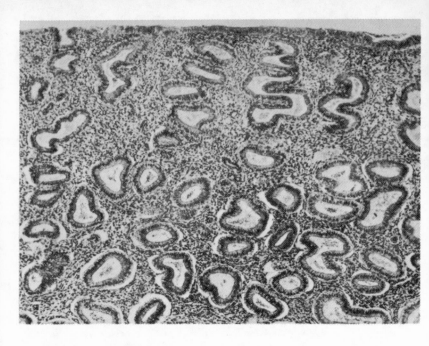

Plate IV. *Sections of the endometrium from a human uterus.* Above: *Just after ovulation.* Below: *Some days later in the cycle.*

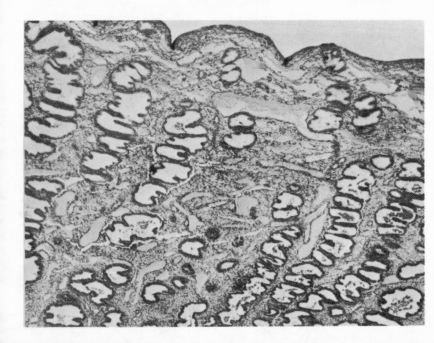

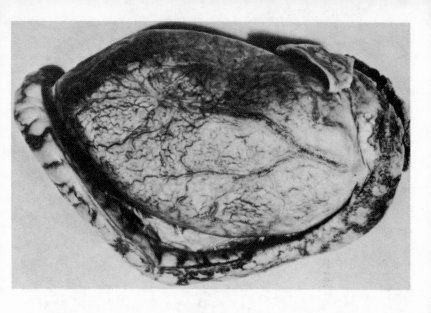

Plate V. Above: *Photograph of the testis of a whale; the epididymis is in the lower part of the picture.* Below: *Section through an active mammalian prostate gland.*

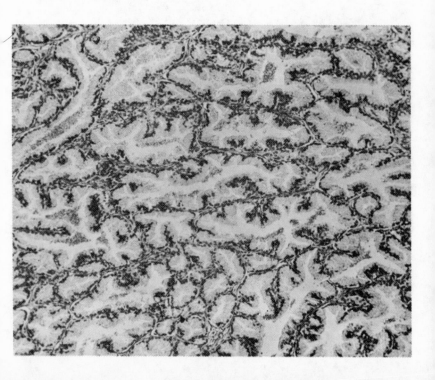

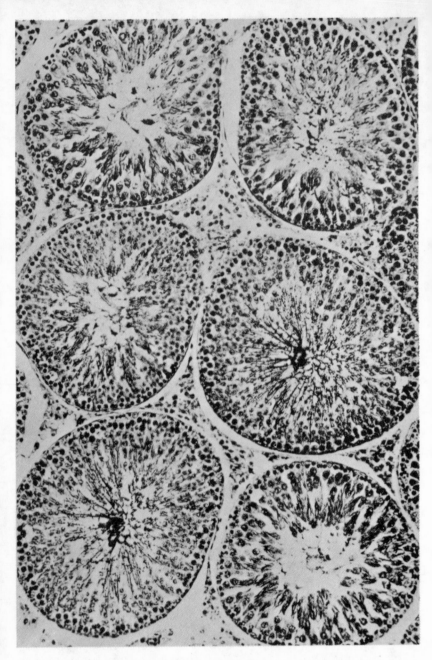

Plate VI. *Section of a mammalian testis showing the seminiferous tubules, the formation of spermatozoa and clumps of interstitial tissue between the tubules.*

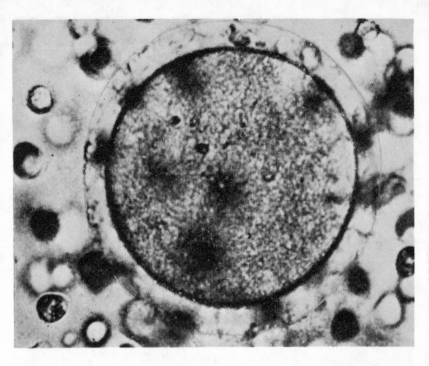

Plate VII. Above: *Photograph of a living human ovum.* (*Photo: courtesy of Professor W. J. Hamilton.*) Below: *Photograph of a preparation of human spermatozoa.*

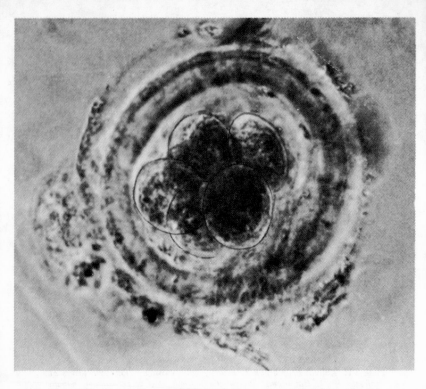

Plate VIII. Above: *Photograph of an early cleavage stage of a rabbit egg seen by phase contrast.* Below: *Photograph of a blastocyst of an elephant seal during delayed implantation.* (*Photo: courtesy of Nigel Bonner.*)

Plate IX. Above: *Photograph of an early human chorionic sac about 25 days old.*
Below: *Photograph of a 28-day-old human embryo.*

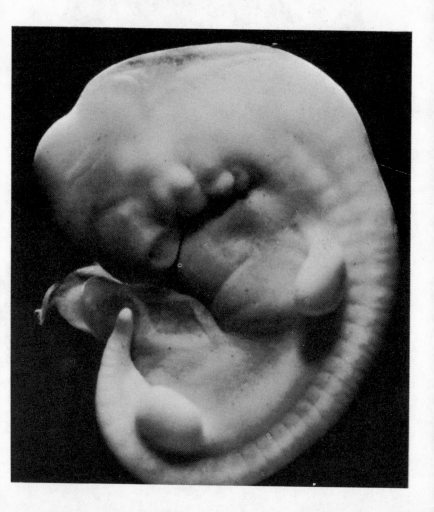

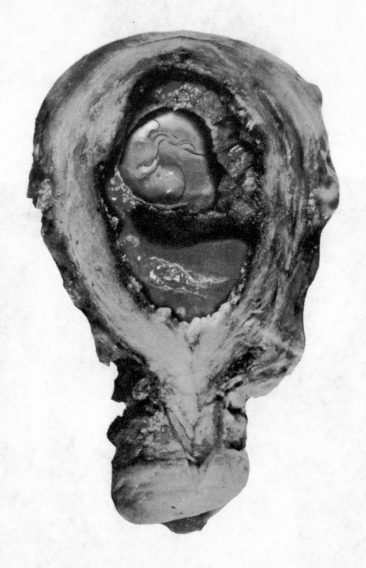

Plate X. *Photograph of an early human pregnancy about eight weeks old. The chorion has been opened to show the foetus lying in the amniotic cavity.*

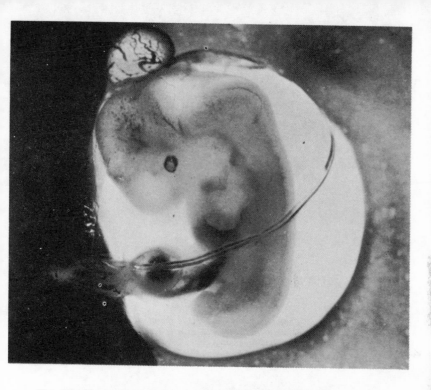

Plate XI. Above: *Photograph of a human embryo six and a half weeks old, 15 mm in length.* Below: *Photograph of a human embryo about eight weeks old.*

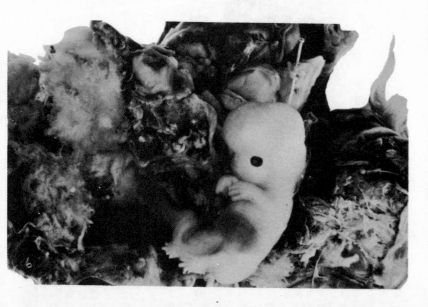

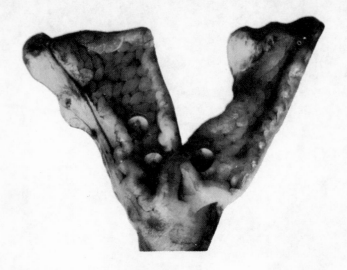

Plate XII. Above: *Photograph of blastocysts in the uterus of a badger during delayed implantation.* Below: *Photograph of a badger foetus and part of its placenta (annular) during mid-pregnancy.*

Plate XIII. Above: *Photograph of the pregnant bicornuate uterus of a rat.* Below: *Photograph of the placenta of a deer (Père David) towards the end of pregnancy.*

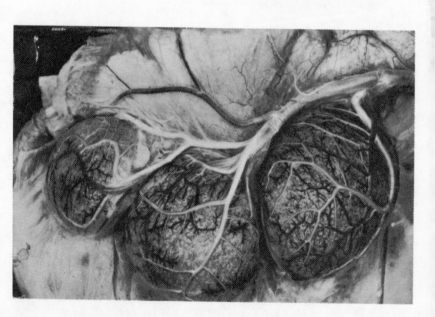

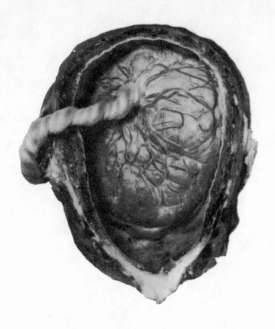

Plate XIV. Above: *Photograph of a human placenta and umbilical cord inside the uterus about mid-pregnancy.* Below: *A cast of the umbilical vessels and placental capillaries in a full-term human placenta.*

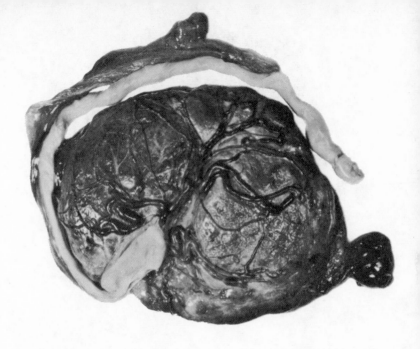

Plate XV. Above: *A shed human placenta and umbilical cord seen from the foetal aspect.* Below: *Photograph of a number of placentomes from a goat uterus.*

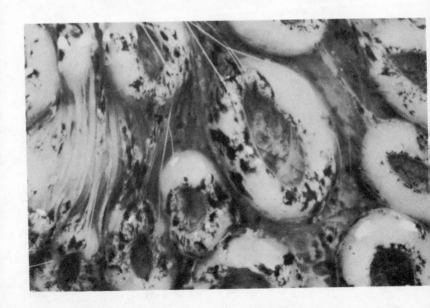

Plate XVI. Above: *A shed human placenta seen from the uterine aspect; note the division into cotyledons.* Below: *Photograph of the uterine aspect of twin human placentae from unlike twins.*

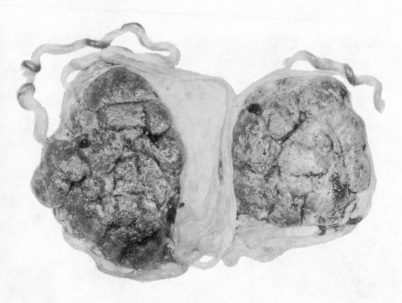

tubules and early spermatogenic activity. Androgenic influences may be necessary to procure the final stages. FSH does not stimulate activity in the Leydig tissue.

The male sex hormones, the androgens, are C_{19} steroids; the two most commonly isolated from the testes are testosterone and androsterone. The Leydig interstitial cells are presumed to be the source of testosterone. Other steroids are produced in males, and in females, by the adrenal cortex. Androgens bring about the development of the male secondary sex characteristics at puberty, and some androgen is necessary to cause maturation of sperm. In females who are given androgens, or in whom an androgen-producing tumour develops in the adrenal, changes typical of males may be produced and there is failure of menstruation. Androgens also cause the sex drive of males and stimulate production of seminal fluid. The production of androgens is brought about by the secretion of interstitial cell stimulating hormone (ICSH). This hormone is identical to LH in females. There are strong suggestions that there is a feedback effect from the testis to the pituitary. Administration of ICSH in immature males causes increased steroid production from the testes and development of secondary sex characteristics. Testicular tissue is also known to produce oestrogens, and they may play some part in regulating pituitary function in males.

As androgen production increases in young males at puberty, the testes grow, the penis enlarges and ejaculation becomes possible. The hair in the axilla and on the pubes becomes abundant and thicker. The hair-line at the temple recedes, a beard develops as well as more hair on the chest and limbs. The voice deepens as the larynx enlarges (Adam's apple). There is increased activity in apocrine sweat glands, and even the mammary glands may enlarge, probably as a result of oestrogen activity. Marked changes in height and weight occur, showing that androgens have generalised effects on the body as well as on reproductive organs. They cause nitrogen retention as well as of salts and water. If the testes are now removed the basophil cells of the anterior pituitary develop vacuoles and other changes characteristic of 'castration' cells.

The pituitary continues to secrete gonadotrophins, mainly FSH, but administration of androgens to the castrate reverses the hormonal and cellular changes. Androgens have been used to 'correct' inadequate sexual development in males, but large doses of androgens suppress pituitary activity and spermatogenesis, and result eventually in testicular atrophy.

Can steroid hormones be used to control fertility in men? Can a 'pill' be developed which will suppress spermatogenesis, as ovulation can be inhibited in females? Oestrogens, androgens and progestins can have a sterilising action when administered in appropriate amounts. Other non-steroid substances affecting testicular activity, nitrofuranes, inhibitors of cell division, antimetabolites and also antibodies to sperm, have been investigated in animals and, to a limited extent, in men. There are, in almost all instances, unwanted side effects such as likelihood of irreversible testicular damage, loss of libido, development of mammary glands and exaggerated responses to taking alcohol. Quite apart from these effects making acceptance most unlikely, there appears to be strong psychological aversion by men to interference with their potential or acknowledged virility. As Gregory Pincus remarks: 'Perhaps experimental studies of fertility control in men should be preceded by a thorough investigation of male attitudes.'

It may well have surprised the reader to learn that a testis produces oestrogens and that ovarian tissue can secrete androgens, but the extent to which they are produced in normal animals and the part they play in reproductive physiology is still obscure. A consideration of the processes of development of the gonads helps us to understand that there is a basic common arrangement of certain structures early on in embryonic life.

According to the nature of the sex chromosomes within the fertilised ovum, so will the young embryo develop in a male or female direction. There is, however, no external indication of the sex of a human embryo during the first eight weeks of its life. There is an initial neutral or indifferent stage during which it is only possible to sex an embryo by examining its cells microscopically for sex chromatin. It has recently become possible to obtain cells of foetal origin during a human pregnancy and so to sex a foetus before birth, but the practical application is limited. Who really would prefer to know the sex of their child before it was born? Even with such knowledge it is too late to change the child's sex.

Embryonic gonads also do not at first show clear signs of whether they are going to develop into testes or ovaries; nor do they at first contain any germ cells. Most embryologists agree that germ cells have an origin outside a mammalian embryo, probably in the wall of the yolk sac. There is some evidence from a study of early human embryos by histochemical methods that there is a migration or invasion of germ cells along the yolk sac stalk into the embryo and eventually into the gonads. They appear to migrate somewhat indiscriminately into organs other than the gonads and even into nervous tissue. It is only in the gonads that they appear to have any chance of survival.

It will be clear that arriving germ cells will be either male or female (XY or XX) and also that cells of the invaded embryo will have similar genetic make-up. The indifferent gonad is induced to become an ovary or a testis as a result of complementary action of its genetic sex and that of the arriving germ cells. The germ cell invasion initiates a progressive divergence in form and type of activity in an embryo towards those of one or other sex. Experiments on embryos of lower vertebrates strongly suggest that removal of the yolk sac early on, and thus of the germ cells, results in failure of sexual differentiation and persistence of a sterile, sexually indifferent stage. Experimental destruction of germ cells in embryos by X-rays also prevents proper differentiation of sexual organs.

Embryonic mammalian testes start to differentiate somewhat earlier than ovaries; a testis is recognisable in a human embryo when it is just seven weeks old. Germ cells steadily increase in number in gonads during foetal life, and by the time of birth each human gonad contains over half a million. What happens from then on in the mammalian gonad has been a subject of considerable controversy. Some have maintained that there is continuous formation of germ cells throughout the reproductive period of life, and in the testis this certainly occurs. There is, however, little evidence that there is a similar process in the ovary. A number of histologists have maintained that they have seen changes in adult mammalian ovaries that suggest neo-formation of oocytes. Experiments on ovaries of rats and monkeys and counts of oocytes in ovaries of mammals of different ages have not confirmed that these changes indicate that new oocytes are being formed. Some additional germ cells may be formed during a few months of the immediate post-natal period. Thereafter it seems likely that no further germ cells are formed and that those present at birth are destined to be those that will mature and will be shed during adult life one by one at successive ovulations. It is difficult to find a germ cell in a post menopausal ovary; thus there must be many that degenerate before reaching maturity.

Not only are gonads indistinguishable during early embryonic life but also external genitalia and those internal duct systems that lead from gonads to the exterior. Embryos of both sexes are initially provided with internal ducts and rudiments of external genitalia which will develop into certain important structures if correctly stimulated. There are four such ducts in any embryo, and their fate is of great reproductive significance. Two are intimately related to the developing urinary system and are known as *mesonephric* or Wolffian ducts (K. F. Wolff, 1733–94, Professor of Anatomy at St Petersburg, described them in the same year that Wolfe captured Quebec). Two other ducts develop alongside them and are therefore called *paramesonephric* ducts. They are often called the Mullerian ducts after J. P. Muller (1801–58), Professor of Anatomy in Berlin. These ducts differentiate into various derivatives depending on the genetic sex of the embryo and how they are stimulated by the developing testis.

It has recently been suggested by A. Jost in Paris that gonads can guide sex duct differentiation by production of substances that act as foetal hormones. The foetal testis exerts an influence over the development of male accessory organs which is independent of pituitary control. Removal of the foetal testes of rats leads to failure of prostatic development and failure of the suppression of the Mullerian ducts that ought to occur in males.

The Wolffian duct in male embryos persists on each side to become the *vas deferens* and *epididymis*. The Mullerian duct of males retrogresses and is of little importance. In females it is the Wolffian duct that degenerates, whereas the Mullerian duct persists. Its upper portion forms the uterine tube of each side; its lower portion fuses with that of the opposite side to a degree depending on the species of mammal. In man, in all Primates except Tarsius, and in bats the lower portions unite completely to form a single-chambered (unicornuate) uterus and the upper part of the vagina. In other mammals the middle portions of the two Mullerian ducts remain separated and their uteri each have two horns (bicornuate). The two horns may unite at their lower ends in a small common chamber opening by a single canal (cervix) into the vagina, as in cattle, sheep, and goats. They may, however, remain quite separate and each horn open into the vagina; this is a primitive condition seen in marsupials and many rodents. Very rarely in man the uterus does not develop properly and a type of uterus seen in other mammals may occur. The significance of a unicornuate type of uterus may be associated with bipedalism. The human foetus lies curled up almost in spherical form within its membranes; a quadruped, on the other hand, has a more elongated foetal form; a different arrangement of uterine muscle is present in the two main types of uterus, with each adapted for expelling the differently shaped conceptus.

The duct systems have developed along one line or the other by the third month of pregnancy, and the external genitalia of the human embryo have also diverged, though not as rapidly. There is at first an indifferent stage in which the external genitalia are represented only by three small simple protuberances. Only a single common chamber draining urinary, genital and alimentary systems opens to the exterior between these protuberances. It is called a *cloaca*, a Latin word for sewer. The mammalian cloaca is soon divided up into an opening at the back for the alimentary canal, the anus, and one in front, the urogenital sinus, receiving the urinary and genital systems. Subsequent developmental changes in this region are complicated and vary in different mammals. They result in the formation in males of a penis and, in most mammals, a scrotum into which testes will descend later. In females the lower end of the fused Mullerian ducts is incorporated between the developing urethra in front and anal canal behind, to form the vagina. The three external protuberances develop into the clitoris, homologous to the penis, and the labia.

Urogenital and alimentary tracts open jointly into a persistent cloaca only in adult monotremes (*mono* = one, *trema* = a slit); a modified cloaca is present in marsupials and some rodents. The penis is found on the ventral wall of the monotreme cloaca, in marsupials it lies behind the scrotum, but in nearly all mammals it lies in front, along the abdominal wall. It is either completely (bull, seal) or partly (Primates) concealed within an invagination of skin, the preputial sac, from which it emerges on erection. Only man, some bats, and tree shrews have a truly pendulous penis with a short prepuce quite free from the abdominal wall.

Migration of testes to a position outside the abdomen occurs only in mammals, but not in all. The testis may not descend at all (whale, elephant), or may descend only into an inguinal or perineal position (seal, rat). It may pass into a scrotum only during the breeding season (bats, some rodents), or, as in most Primates and man, it descends into the scrotum just before birth. Various factors control the descent of the testis (a thin cord, the gubernaculum, guides it more or less on its way), but it is still difficult to explain these differences amongst mammals. Evidence has been produced which suggests that in man and many mammals the scrotum has a heat regulating function and that the male germ cells can only mature into spermatozoa at a temperature lower than that in the abdomen.

The scrotal temperature is certainly a little lower than the abdominal in man, but the habit of wearing clothes hardly seems to reduce the potency of males. Undescended human testes do not start functioning effectively at puberty unless brought down previously by operation. Endocrine activity is mildly damaged, but spermatogenesis considerably and even permanently. We have no knowledge of what would happen to an elephant's testis were it to descend into a scrotum!

SUMMARY

The male gonads, the testes, produce spermatozoa from convoluted seminiferous tubules and also male sex hormones, androgens, from groups of interstitial cells lying between the tubules. The testes start developing within the abdomen and in a majority of mammals (except cetaceans, pinnipeds, sirenians and elephants) 'descend' through the inguinal canal into a scrotal pouch. Failure to 'descend' in those species in which it should results in an inactive testis.

The testis becomes functional at puberty as the result of pituitary stimulation. Spermatocytes in the epithelium of the seminiferous tubules divide into spermatids. These are nourished by Sertoli cells as they differentiate into spermatozoa which are over 50 μ in length and have a head, containing the chromosomes, and a long propulsive tail. The spermatozoa pass from the testis by collecting ducts and mature in the coils of the epididymis. The prostate, an organ surrounding the first part of the urethra, and the two seminal vesicles at the base of the bladder, provide the seminal fluid. At ejaculation in man some 45–700 million spermatozoa are expelled in from 1·0–10·0 ml of seminal fluid. The fluid provides nutriment and an appropriate environment for the spermatozoa, it stimulates them to become motile and acts as a vehicle for their conveyance to the female reproductive tract. In the right circumstances spermatozoa can remain viable for up to 72 hours in the female genital tract. Spermatozoa are propelled by the undulatory movements of their tails, and each progresses with an oscillatory motion along a spiral path. Mammalian sperm are remarkably similar in their morphology, with only minute differences in their heads but with, of course, quite different genetic instructions in their chromosomes.

The human male is a 'constant breeder' in that from puberty until quite late in life the testes produce spermatozoa continuously. So also do the testes of most Primates and rodents; but males of many other species exhibit a season of testicular inactivity or decline. When a stag, or a billy goat, is in season it is in rut and may be violent and very aggressive. The interstitial cells (forming the Leydig tissue) of the testis are the source of male sex hormones, androgens. These hormones are produced in increasing quantities before and during puberty, and they cause the development of the male secondary sexual characteristics. If the testes are removed before puberty, the secondary characteristics do not develop and a eunuchoid state appears in which an adolescent build is retained in the adult.

The determination of sex of a new individual is effected at fertilisation by the chromosomes in the ovum $(22 + X)$ being joined by those carried by a spermatozoon. Should the haploid number of chromosomes in the fertilising spermatozoon be $22 + X$, then the zygote will have a diploid number of $44 + 2X$ and be *female*. Should the spermatozoon contain $22 + Y$, then the reconstituted number is $44 + X + Y$ and the new individual will be *male*. Alterations in the diploid number, due possibly to faulty chomosome number in the germ cells, are associated with certain abnormal conditions.

According to the determination of sex, so certain duct systems become stimulated in an embryo to develop into the secondary reproductive organs. The male state results in the persistence and development on each side of the mesonephric (Wolffian) duct into a vas deferens and epididymis. In females this duct (most of it) retrogresses, and the paramesonephric (Mullerian) ducts persist and become part of the vagina and all of the uterus and uterine tubes. The Mullerian ducts of males retrogress. It is thought that in some mammals at least the foetal testes influence the development of the duct systems. As the ducts differentiate, so do the external genitalia: at first both male and female embryos look alike, during a neutral or indifferent stage. There is a common cloaca which receives the lower ends of the developing alimentary canal, urinary tract and genital systems. Small genital protuberances surround the cloacal opening. Growth changes result in a separation of the anal canal from the urogenital sinus. Later, the protuberances give rise to the labia and clitoris of females, or to the penis and scrotum of males. Failure of these developmental changes can result in types of pseudo-hermaphrodite and other abnormalities.

5. Early Stages in Pregnancy

There has always been some difficulty over definitions and terms in the early stages of pregnancy. When is an animal or a woman really pregnant? Is it at fertilisation of the egg, or at implantation in the uterus? Does fertilisation really denote the creation of a new soul? In a way, many of the terms are derived *post hoc ergo propter hoc*. The final event means that the earlier must have occurred; and when the earlier failed, the final one never could become manifest, so perhaps the earlier never did occur. How many of us became potential mothers or fathers, or temporarily had twin brothers? The answer is, more often than we would have thought: about two to three times more often.

Fertilisation is the series of processes by which a spermatozoon initiates and participates in development. It starts with approach of spermatozoa to an egg and concludes with fusion of the pronucleus of the egg with that of one spermatozoon. Many factors are now recognised as being intimately concerned with the process. Although Leeuwenhoek described spermatozoa in 1679, and von Baer first saw a mammalian ovum in 1827, we still know little about fertilisation in man. We do not know whether human eggs, as do some eggs, produce substances that increase sperm motility or attract them. We do know, however, that fertilisation is an example of a tissue-specific reaction in that sperm will enter only eggs and in that eggs will receive only sperm. In general, spermatozoa will not penetrate eggs of other species. It is not therefore surprising that theories of fertilisation are based on analogies with immunological responses.

Extracts of sperm of various species have the property of breaking down egg membranes. Lytic agents in mammalian

74

sperm extracts cause dispersal of the follicle cells surrounding unfertilised eggs by dissolving their close contact. This property is probably due to presence of an enzyme hyaluronidase, so called because it can break down the mucopolysaccharide hyaluronic acid. Recent research has suggested that in certain mammalian species at least an individual spermatozoon may carry enough enzyme to make a path for itself through follicle cells about an egg. In man, spermatozoa probably meet the egg in the outer third of the uterine tube. Several may reach the zona and even penetrate it; only one enters the oocyte (vitellus) and can do so anywhere on its surface. The head immediately swells and becomes the male pronucleus; the middle piece and tail are lost and absorbed.

The oocyte now undergoes its final maturation division, sheds another polar body, and the remaining chromosomes form the female pronucleus. The whole egg cytoplasm, or vitellus, shrinks within the zona, and in some species a 'fertilisation membrane' is formed on the surface of the vitellus. Electron microscope studies suggest that this may be a result of re-orientation or even extrusion of fatty materials on the surface of the vitellus. Changes in the oocyte, perhaps the creation of a fertilisation membrane, prevent further spermatozoa penetrating the vitellus. The two pronuclei meet near the centre of the human ovum and a one-cell fertilised egg, or zygote, is formed. It only remains a single cell for a few hours: the stimulus of fertilisation soon brings about a series of divisions into two, four, eight cells. This process is called *cleavage*.

PARTHENOGENESIS

When eggs are activated without intervention of spermatozoa the individuals that subsequently develop are said to do so parthenogenetically (*parthenos*, a virgin). The process plays an important part in the life story of many invertebrates. Generations produced in this way may alternate with those resulting from normal fertilisation. Various types of stimulus can experimentally activate eggs of invertebrates and even lower vertebrates; parthenogenesis has not, however, been convincingly demonstrated in normal mammalian life. Mammalian eggs (rabbit) can be artificially activated to divide through a few cleavage stages, but have to be transplanted into a receptive uterus for further development to occur.

SEX RATIOS

The ratio of the number of boys to that of girls born in any period is called the *secondary* sex ratio. The Registrar-General's *Annual Reports* give the ratio for Great Britain in each year; it is now 105·6:100. This means that more boys are born than girls, but there is evidence that a still greater proportion of eggs are fertilised by Y-containing sperm. This is the *primary* sex ratio. It is difficult to determine in man, but in pigs it is about 160:100. The sex ratio of human embryos and foetuses that die before birth for one reason or another is high (one set of observations gives a ratio of 150:100). Numerical preponderance of males continues through early age groups; equality is reached by the age group 15–19 years, thereafter a female ascendancy develops until at age 85 there are twice as many females as males. The state of 'maleness' carries with it a certain fragility.

Many hypotheses have been put forward to explain the sex ratio and this differential viability of the sexes. It has been frequently maintained that those sperm with a small Y chromosome can swim faster, or are more active, than the supposedly heavier X-carrying sperm. Y-carrying sperm have been suggested to have a higher metabolic rate, as does the male diploid generation of all species studied. Recently efforts have been made to ordain the sex of calves at will by attempting to separate the two kinds of sperm before artificial insemination. The results were unsuccessful. A certain type of environment within the female genital tract may favour one or other type of sperm. Sex-linked, or sex-limited, recessive lethal genes may be responsible for some pre-natal deaths of male heterogametic foetuses. Heterozygous-defective genes may accumulate in males because of a less intense selection than in females. The secondary sex ratio appears to be affected by many factors; they include urbanisation, social upheaval, whether first-born, migration, whether illegitimate, social class, and age of mother. Few serious statistical studies have been made on such matters. A thorough study by A. Ciocco in 1938 in the United States found little evidence of much influence on sex ratio by such factors, and he condemned speculation on their significance.

In a light-hearted vein it may be remembered that it has been suggested in the past that the sex of a child was determined by which testis was active or by the side on which the woman lay at conception, and that overall in a family the children would be the *opposite* sex to the dominant partner in the marriage. There is, however, some slight (but not, regrettably, to be relied on!) statistical indication that if a family has been all of the same sex and born within a short space, a pause in the programme may result in the birth of a child of the opposite sex.

SIGNIFICANCE OF FERTILISATION AND CLEAVAGE

Fertilisation marks the embryological origin of a new individual. The diploid number of chromosomes is restored, sex is determined, and cleavage is initiated. The restored chromosome number and related genes determines the characters of the species of embryo (*genotype*). As the embryo develops, effects of environment, both pre- and post-natal, will influence the organism so that it appears as a *phenotype*. We are all human beings, but we all show phenotypic variation: peculiarities of heredity in man are due in many ways to his capacity for outbreeding. Inbreeding is virtually limited to that of race and class; his almost instinctive objection to incest has undoubtedly been to man's immense advantage.

The genetics of sex determination have been described on p. 61, but every now and again a child is born in which it is difficult to ascertain sex. Such imperfections in sexual anatomical development can be loosely called 'intersexes'. They suggest that it is not just presence of sex chromosomes, but the quantity of X and its ratio to that of autosomes that finally determines sex. Occasionally an individual thought to be a girl at birth is brought up as a girl but is really a boy with poorly developed male characteristics. Intersexes occurring in man include very rare individuals called *hermaphrodites*, of whom only about 100 have been described. They possess either one ovary and one testis, or an ovo-testis on one or both sides of the body; usually only one type of gonad is active. There are also a few families in whom the X chromosome of individuals apparently of the XY category is altered by presence of another gene. These individuals ought to be males but they turn into apparent females. Externally their anatomy is female, but they possess no uterus and menstruation occurs only from monthly nose-bleeding. Ovaries have been transformed into poorly developed sterile testes. A *pseudo-hermaphrodite* possesses gonads of one sex and external genitalia of the other, either wholly or, more likely, partially developed. They are commoner than hermaphrodites, and the incidence, including minor aberrations, is about one in every 1000 individuals; many authorities believe it to be even higher. Other, but rarer, chromosomal aberrations are recognised. They include a syndrome (Turner's) in one individual of infantilism, webbing of the neck and inwardly deviated forearms; there may be faulty gonadal development. The chromosomal constitution is 45 with XO sex chromosomes. They could therefore be females who have lost an X-chromosome, or males who have lost a Y-chromosome. Another syndrome (Klinefelter's) is associated with atrophy of the seminiferous tubules, lack of sperm and some mammary development. The sex chromosomes are usually XXY. Examples are also known of individuals with XXX and XXXY, and of individuals who are compounded of cells of two or more chromosomal types. Mosaics of several kinds are known, such as XXY/XX and XO/XX. One probably unique mosaic of XO/XXX exhibited certain poorly developed characteristics of both sexes. One explanation of these findings is that of 'non-disjunction', by which is meant a transmission of both X-chromosomes (or neither) into the definitive egg nucleus as a result of faulty gametogenesis. The number and type of chromosomes may be determined in human beings in various ways. Essentially the techniques involve the utilisation of dividing cells obtained by culturing bone marrow, skin or leucocytes. The cells are fixed, stained and squashed out flat, often after pre-treatment with a drug that inhibits spindle formation. A photograph is made of the chromosomal array, and the individual chromosomes are then cut out, arranged in pairs according to an agreed scheme (Denver) and rephotographed. Such methods, together with those of chromatin sexing (p. 64), have provided much information on the genetic constitution of intersexes.

Cleavage can be defined as a series of rapid divisions in the zygote. In human miolecithal eggs the whole egg divides into

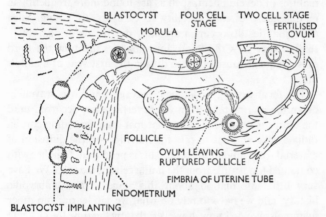

Fig. 12. *Diagram to show the events following ovulation and fertilisation and leading up to implantation of the blastocyst.*

two equal-sized parts which again subdivide equally (complete and equal cleavage, as opposed to birds' eggs in which abundant yolk causes cleavage to be partial and incomplete). An increasing number of cells (blastomeres) are produced and, as there is little increase in total cytoplasm during cleavage, there is reduction in cell size to that of the body cells. Material in the zygote is partitioned among the blastomeres; even in mammalian zygotes there is evidence of uneven distribution of certain substances. Cleavage marks the first stage of differentiation at a chemical level in embryonic cells. It also allows a certain degree of movement of cells within the zona. Groups of cells eventually become arranged in different ways, so that there is a change in shape as a result of morphogenetic movements. Histogenetic and morphogenetic developments go hand in hand as an embryo grows; as their names suggest, we believe the processes to be under genetic control. It will be obvious that they could suffer interference by outside factors or by adverse conditions developing within embryonic cell systems. Effects of such interference could be drastic, even lethal, or of a more subtle nature. They could lead to abnormalities of development, such as are found more frequently in man because of man's ability to care for his young. Such abnormalities do occur in animals, but the deformed young seldom survive. Congenital abnormalities are now the third most frequent cause of death during the first year of a child's life.

The fertilised egg undergoes cleavage while it is travelling down the uterine tube from where fertilisation occurred until the uterus is reached. The egg has lost its surrounding corona radiata cells and has to rely for nutrition on the diffusion of material from fluid in the tube. This is produced, together with contained foodstuffs, from the lining cells of the tube. We have very little information about this stage in our embryonic history, and we have to rely on what we can learn from other mammals. We do know, however, that by the time the cleaving egg has reached the uterus it is 4–5 days old, has divided into a hollow sphere of cells and is now referred to as a free *blastocyst*.

The free blastocyst is small at first in all mammals; it is 0·09 to 0·1 mm in diameter in man. It consists of a hollow sphere of cells one layer thick, with an *embryonic knot* inside at one pole. Four and a half days after fertilisation a human blastocyst contains just over 100 cells. Even at this early stage one can detect a division of labour taking place amongst its cells. Its single-layered covering is called the *trophoblast*, and is destined to give rise to membranes and placenta that surround and nourish the growing embryo. The embryonic knot (inner cell mass) will develop steadily into an embryo.

The young blastocyst is nourished by diffusion of foodstuffs through the trophoblast. In mammals it reaches the cavity of the uterus between the fourth and ninth day after ovulation; it arrives there on the fourth to fifth day in man. The corpus luteum in the ovary has by then reached its fullest activity and caused transformation of the uterine lining, so that the latter secretes material (uterine milk) nutritious to the blastocyst. The zona pellucida (p. 33) usually ruptures and is lost when the blastocyst reaches the uterus: the blastocyst can now 'implant'.

IMPLANTATION

This term signifies a process of attachment of the trophoblast to the maternal uterine tissues. Several types of implantation are recognised, but all depend on an initial contact being made between trophoblast cells and uterine epithelial lining. What follows depends on the species of mammal. In pigs, sheep, cows, horses, dogs and cats the blastocyst enlarges rapidly and expands so much that it soon fills the greater part of the inside of the uterus. The area of contact between expanding trophoblast and uterine lining increases steadily. This method of implantation is called *central*, or *superficial*. In man, chimpanzees, and certain bats the blastocyst becomes embedded in the uterine lining – *complete interstitial* implantation. This occurs partly as a result of 'burrowing in' and partly by enclosure within the uterine lining. In a third group of mammals that includes rats and mice, the blastocyst enters a recess in the uterine cavity in which it becomes closed off and where it implants. This is known as *eccentric* implantation but can be considered to be a modified type of interstitial implantation.

Blastocyst attachment occurs in man 4½–7 days after fertilisation. No normal examples of this stage have so far been recovered. Implantation proper begins 5½–6 days after fertilisation, when the blastocyst begins to embed in the uterine lining. Glandular activity increases, and as soon as implantation occurs there is marked increase in blood flow and intracellular accumulation of fluid (oedema). These changes lead rapidly to those of a 'decidual reaction'. This involves enlargement of connective tissue cells with glycogen and fat.

There are also changes in intercellular substances that probably make the lining more resistant to invading trophoblast. The changes serve the twofold purpose of providing nutriment for the young embryo and also of containing invasive propensities of trophoblast cells. They may also prevent excessive haemorrhage from ruptured maternal vessels that might dislodge the embryo. The reaction of the lining therefore prevents too deep initial penetration, thus ensuring both successful implantation and survival of the young embryo.

The mechanisms controlling implantation are believed to be very complex and to involve many factors. Much experimental work has attempted to elucidate the action of these factors and even to control the process. Besides the many workers investigating these problems all over the world, there is a whole department in Israel whose entire staff are devoted to solving what happens at implantation. Several general statements can be accepted. It is clear that the phenomenon is not a casual or random one, because both the egg and the uterine lining must have passed through various phases of development before implantation can occur. There is a mutual interaction between the embryonic and maternal tissues, the details of which appear to differ among mammalian forms. Various phases in the mechanism can be broken down and shown to have at least some hormonal or chemical control. The uterine lining must have reached a certain stage in its 'ripening'. In some animals an experimentally induced reaction can be procured in the uterus by the introduction of glass beads or wax pellets into the uterine cavity at a particular time in the cycle. It is also possible to culture pieces of the uterine lining cut out from the uterus and then artificially to implant blastocysts on the cultured mucosa, though the blastocysts do not live for long. The blastocyst has also to reach a certain stage before it will even attach to the uterine lining, and the zona pellucida must have dissolved or become detached. Many workers believe hormones may cause the changes in the zona and suspect that local action by progesterone or oestrogen may be responsible. Certainly in some forms local injection of a 'surge' dose of oestrogen can precipitate implantation under certain conditions. There are also suggestions of specific tissue reactions and liberation of chemical substances in the uterine mucosa at the site of implantation. The physiologically active substance histamine has been thought to be involved in the tissue reactions and to have been caused to be excreted from certain cells by the arrival of some chemical 'signalling' substance from the blastocyst. Recently attempts have been made to test complex chemicals which might have a toxic effect on the blastocyst and not affect the mother. This deliberate attempt to *kill* the products of conception may not seem as desirable a form of contraception as methods designed to *prevent* liberation of an egg from the ovary, or to *prevent* the germ cells meeting, but it is really a chemical form of abortion. So far, such substances are only known to work (and not always) in rats.

It seems surprising that, with the mechanism of implantation apparently so complex, so many successful implantations occur, some indeed that were never intended. It is, however, a mechanism that has been evolved and has become effective over 50 million years, and it is both a preliminary and an essential stage in viviparity. We know, too, that it is not always successful and that failure in implantation is not uncommon both in mammals and man. Careful examination of some of the implanting blastocysts which have been recovered, and at first thought to be normal, shows them to be 'blighted' and doomed either to extrusion from the uterine lining, or to absorption or to loss at the next menstrual flow. Perhaps these blastocysts were themselves abnormal, incomplete or contained lethal genes, or perhaps the uterus had failed to become adequately prepared for their reception, or did not respond properly when it should have done. Whatever the cause, these failures are no real loss, and they indicate some degree of natural refusal by the implantation mechanism of faulty components. Again, the mechanism does not always exhibit a similar pattern in all mammals. There appear to have been certain alterations in the timing of the phases in implantation which became incorporated in the behaviour of some species. They exhibit a *delay* in the process, and its existence may be a guide not only to how implantation occurs but how to control it.

Delayed implantation

This curious phenomenon occurs in a number of apparently
unrelated mammalian groups. They include marsupials,
armadillos, martens, badgers, seals, sea lions and roe deer.
The blastocyst reaches the uterine cavity in the normal way,
but from then on there is a pause in development that can last
many months and in badgers even up to two years. The
blastocyst lies dormant, in an almost hibernatory state, in a
recess in the uterus. The uterine lining undergoes little secretory change during the period of delay until, in response to
some ill-understood stimulus, implantation suddenly occurs at
a particular time of year and development continues normally.
It has often been maintained that a hormonal mechanism is
responsible for delay. Badgers, however, can ovulate more
than once even while blastocysts have been present in the
uterus for some months. It is also tempting to suggest from
their appearance that the zona pellucida of these blastocysts is
more resistant than that of other mammals, and so prolongs
the free blastocyst stage, but there is no definite proof.
Delayed implantation can result in birth of young at the same
time every year, often at a time most favourable for the
newborn. It also allows the females of a species to give birth
and to mate almost immediately afterwards, then for the sexes
to migrate widely and in different directions, and then, after a
period of delay and of normal pregnancy, to return to the
breeding grounds at the right time of the year. Delayed implantation can thus be advantageous to both young *and* parents,
and to a species.

Another form of delay occurs in some rodents. The female
gives birth to many young and suckles them. A few days after
birth, ovulation and mating occur (known as a post-parturient
ovulation). Fertilisation takes place, and blastocysts reach the
uterus. Then, depending on the number of young being
suckled over the first three, there will be that number of days
delay in implantation (usually never more than nine). This type
of delay is, therefore, related to hormonal thresholds during
lactation.

True delay in implantation, or even that related to lactation, has never been reported in any Primate, including man. There have, indeed, been occasions when the time of birth has been a month later than expected, but there are many other and more practical explanations for these late arrivals than ascribing them to delay in implantation. One difficulty we do have to consider is whether delayed implantation is a primitive characteristic, possessed at one time long ago much more widely among mammals. The passage of time, changes in climate, adaptation and plasticity in the reproductive pattern, may all have resulted in its loss in most mammals but its retention in the few now displaying it. We leave it to the reader to decide whether delayed implantation would be an advantage to man. It would, however, be a worrying matter to be all the time uncertain whether a change of climate on holiday or an aperitif in the sun might precipitate an addition to the family!

EARLY HUMAN DEVELOPMENT

The human blastocyst retains an almost solid form during the first phase in implantation; the outer trophoblast becomes two-layered during the second phase; by the third phase the blastocyst has become hollowed out by certain cavities. The trophoblast divides into an outer *syncytio*trophoblast (so called because it seems to be a multinucleated layer lacking individual cell membranes) and an inner *cyto*trophoblast (often called after Theodor Langhans, who described it in 1870). Both layers produce hormones, one of which it is suggested has an important function relating to a fate that threatens the young embryo – its expulsion from the uterus at the menstruation that should shortly take place. The menstrual flow that would be expected some 14 days after ovulation is in fact suppressed. There is evidence from several sources that it is a hormone from the cytotrophoblast (a chorionic gona-dotrophin) that acts on the maternal ovary of pituitary, or both organs, to cause menstrual suppression. This hormone is soon excreted in biologically detectable quantities in maternal urine; methods have been devised for demonstrating its presence, and form valuable pregnancy tests. It is not known for certain that the trophoblast of any other mammal produces a gonadotrophin comparable to that of man, although it is too early to call it a unique human characteristic. There is evidence that the trophoblast of the mare's blastocyst may stimulate the endometrium to produce a comparable hormone. We are not yet quite sure precisely how embryos of every mammalian species weather early hormonal storms and other inclemencies.

The syncytiotrophoblast possesses enzymes capable of destroying maternal tissue with which it comes in contact. Some other factor may also cause the death and liquefaction of maternal cells that occur close to the trophoblast. Tissue death results in formation of material called *histiotrophe* (nourishment by tissue) that is nutritious to the embryo. Implantation is well advanced by the 12th day after fertilisation, and the embryo is veritably a parasite within its mother, though, as we have seen and shall see again, the placenta plays an important part in controlling the hormonal climate of pregnancy.

Small projections develop on the outside of the trophoblast; these are primitive *villi*. Their appearance increases the surface absorptive area of the trophoblast, and they develop into the placenta, becoming complicated in form and gaining blood vessels that link up with those of the embryo (Fig. 13).

The embryonic knot soon becomes transformed into a flattened, pear-shaped disc. This is brought about by appearance of two cavities within the blastocyst; details of their formation differ in various mammals. They are the *amniotic* and *yolk sac* cavities. Between them lies the germinal disc composed of two layers of cells. The layer forming the floor of the amniotic cavity is called the *ectodermal* layer; that on the under-aspect of the disc is the *endodermal* layer. They are the first of what von Baer christened *germ layers*: they are found in all vertebrate embryos. A third layer, the *mesodermal*, soon appears between the other two. It is technically known as *secondary* mesoderm, to distinguish it from *primary* mesodermal cells that take part in formation of foetal membranes.

If we could look down into the amniotic cavity of a human

embryo about 15–16 days old, we should see below us the upper ectodermal surface of the pear-shaped germinal disc; its broad end marks the front or head end. A mid-line longitudinal furrow that extends along most of the disc is the neural groove. It is destined to develop into neural tissue of brain and spinal cord. A small node is present at the hind end of the neural groove; it was first described by a German embryologist, Viktor Hensen. Behind Hensen's node a mid-line streak marks the site of formation of secondary mesoderm below the surface ectoderm. Study of vertebrate embryos suggests that cells formed here migrate forwards to provide embryonic connective tissue.

A column of cells originates from a point close to Hensen's node and grows forwards beneath the neural groove towards the head end of the disc. This is the *notochordal* process, a slender, mid-line rod of cells which clearly divides the disc into a left and a right side. It indicates a future bilateral symmetry shown by all vertebrate embryos and marks the earliest stage in development of a backbone. Possession of a notochord some time or other in life by an animal means that it can be classified as a *Chordate*. In lower, free-swimming Chordates, such as *Amphioxus*, the notochord remains as a long, flexible, and elastic rod, and its main function can be compared to that of a strut preventing the animal from being telescoped. The largest sub-phylum of the Chordata is the Vertebrata, and in them the notochord is absorbed into and replaced by the spinal column.

The secondary mesoderm formed from the primitive streak and elsewhere lines up on each side of the notochord and becomes segmented into a series of blocks called *somites*. Segmentation starts near the head end and continues towards the tail end until eventually some 44 somites are formed. It is their presence which allows us to speak of a 20-somite embryo – or of a presomite embryo before they have developed. The significance of the somites is that they illustrate the basic construction of much of a vertebrate embryo. Somitic segmentation (metamerism) primarily involves the muscular system but also affects other features. It is reflected in the serial arrangement in mammals of ribs, vertebrae, intercostal nerves, and skin segments of the trunk. Much modification of basic segmental pattern has occurred in other parts of the body. Knowledge of details of modifications in the basic segmental pattern can be very helpful in understanding adult structure and in elucidating how certain abnormalities and disease processes affect the human body.

A second important change in somitic mesoderm involves formation of a split in the outer edges of each somite, so that it becomes divided like a sandwich. The upper layer will necessarily come to lie on the inner aspect of the ectoderm; the lower layer will become applied to the endoderm. There is thus a cavity left between the two layers on each side of the germinal disc. It is to become the *coelom* which is later subdivided into important body cavities. All the subdivisions are at first in communication in a young embryo. Later they become anatomically separated and, apart from certain developmental abnormalities, remain so throughout life. Separation of the subdivisions is related to development of folding of the germinal disc.

Folding of the early embryo

Folding, or 'tucking under', now occurs in the head and tail regions and at the sides of the germinal disc. Mesoderm, called septum transversum, and the primitive heart tubes first develop in the head region. Folding of the embryo results in these structures coming to lie beneath the head: the septum transversum cuts across the coelom as the diaphragm. It thus helps divide pleural from peritoneal cavities. Folding in the tail region provides room for development of openings for alimentary and urogenital systems, and for a tail in those species that have one. Folding at the sides of the disc encloses the digestive tube and viscera that will develop from the endodermally lined yolk sac. The inner layer of mesoderm (see above) of each side closes round the gut tube to provide its muscle and connective tissue. The outer layer, together with ectoderm, closes in front of the now almost tubular embryo to form the tissues of the anterior thoracic and abdominal walls. For some time, however, the developing intestines remain outside the embryonic body. Should the body wall not complete properly, the baby is born with its intestine protruding at the umbilical region.

Folding processes result in the young embryo 'rising up' in the anmiotic cavity. The amnion ceases to be attached to the edges of the disc, and is swept round until it surrounds a stalk left between the embryo and the developing placenta – the stalk is the *umbilical cord*. It is possessed by all Eutherian mammals and is evidence of true intra-uterine life with a placental attachment. The embryo elongates more rapidly than the amniotic cavity enlarges in diameter: thus it develops a marked curvature and is 'doubled up'. Young embryos are thus measured in a straight line from crown to rump – thus the expression C.R. length.

Folding results in the heart coming to lie below the head, with the front, narrowed part of the yolk sac (fore-gut) extending between them. A number of interesting pharyngeal arches develop on each side of this primitive mouth cavity. In fishes, salamanders and tadpoles they are called *gill bars* and have slits between them that permit passage of water. Gills have a respiratory function in these animals; oxygen passes from water to blood within their vessels. Pharyngeal arches or bars do not have a respiratory function in reptiles, birds or mammals, nor are there slits between the arches. In mammalian embryos derivatives of pharyngeal arches contribute to structures in face and jaws, ears, tongue, larynx, and to several endocrine organs in the neck. Each pharyngeal arch has ectoderm outside, an endodermal lining and cartilage, muscle, blood vessels, and nerves within it. Analysis of the fates of the various arch components is essential for understanding adult anatomy in the head and neck. It also throws much light on evolution of jaws, auditory mechanisms, and endocrine organs.

Temporary existence of pharyngeal arches has frequently been used to support theories of recapitulation, i.e. 'that embryos of higher forms pass through stages resembling adults of lower forms' (von Baer). Such theories are now only

R.M.—5

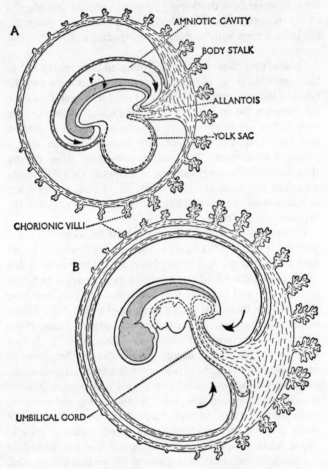

Fig. 13. *Diagram of the chorionic sac, foetal membranes and embryo in early stages of human development. Folding movements result in the formation of the umbilical cord.*

considered helpful after some modification. Pharyngeal bars in mammals are not gills: they represent a structural groundwork occurring in vertebrate embryos which are used in some to build gills and in others to make jaws and other structures.

We may safely deduce from the similarity of this arrangement in all classes of vertebrates that 'they inherit their pattern of growth through some one common protovertebrate ancestor, of fish-like general character though not exactly like any present-day fish. The descendants, as they evolved into various classes, necessarily inherited such a pattern, but they worked it over into new forms suitable for their needs' (G. W. Corner).

Studying human embryos

It is difficult to observe a human embryo as it develops within what Sir Thomas Browne called 'that other World, the truest Microcosm, the Womb of our Mother', and it is virtually impossible to carry out anything in the nature of deliberate experiments. We have to rely for information on how the human embryo develops by examination of embryos that are accidentally disrupted from the maternal uterus – an event called an abortion if it occurs before the end of the seventh month of pregnancy, or premature labour if it occurs after that date. The difference is really only a legal one, because the foetus is considered at the 28th week of pregnancy to have reached a 'viable' stage. This implies that legally it is a citizen of this country and thus capable of enjoying certain rights of its own.

Human embryos occasionally become available to the embryologist as the result of surgical termination of pregnancy in order to save the mother's life, or to ensure that her health would not be permanently and adversely affected were the pregnancy allowed to continue to term. Procedures to procure abortion have at times been permitted in some countries for reasons other than those of health, but in Britain the law is strict and can still only relent for good medical reasons. Whether or not the laws will have to be relaxed as over-crowding develops from increasing population is, at the moment, only an academic consideration, but it may not always be so.

Human embryology is a young science; only in 1942 was an embryo as early as $7\frac{1}{2}$ days old obtained. This 'veritable jewel

in the treasury of science', as the Director of the Carnegie
Institute of Embryology described it, was recovered by Drs A.
T. Hertig and J. Rock of Boston. Up to the time these two
men began so extensively to increase our knowledge, the
stages in early development of human embryos had to be
inferred from studies of early pregnancy in Primates.

A most important reason for knowing all we can about the
formation of embryos and their organs is that certain drugs,
chemicals, physical factors such as radiation, and even naturally occurring substances such as vitamins and hormones can
interfere with the mechanisms controlling the process.
Abnormal organs can develop, or the normal process can fail
to complete properly, or organs and parts may not appear at
all. There are a number of imperceptibly overlapping phases or
stages in human embryonic development. The factors affecting embryonic growth are so many, however, that each
embryo cannot be expected to follow a distinct and exact
time-table. Differences between embryos may not be very
striking in the early stages, but later in pregnancy they may be
sufficient to cause marked divergence in size and weight of two
foetuses of the same age.

Horizons in development

An eminent American embryologist, G. L. Streeter, thought it
might be advantageous to borrow a term used in geology in
order to describe human embryological events. He took the
word 'horizon' and used it, as in its geological sense, to imply
an epoch or period of time covering a number of events. He
described only eight of his suggested 23 horizons for the early
period of human development before he died, but others have
been able to continue his work. Descriptions of the successive
horizons were based on the magnificent collection of early
human embryos housed at the Carnegie Institute of Embryology in Baltimore, a collection which is unequalled in any other
part of the world. The technicalities of the descriptions are
only for expert embryologists, but the general conclusions
provide us with an excellent review of early human developmental stages. A number of horizons will be grouped together

here, for so much happens so quickly in the early weeks of human embryonic life. It is hoped that another volume in this series will deal more fully with the growth of the human embryo and foetus.

It is essential to realise that each horizon in development is not a fixed or static stage. Every geologist finds particular fossils or distinguishing features in his 'geological horizons', yet he does not imagine that they mark the end of a stage in evolution. Streeter remarked that it is necessary to emphasise the importance of thinking about an embryo as a living organism 'which in its time takes on many guises, always progressing from the smaller and simplest to the larger and more complex'. The early horizons are eleven in number and cover the first 24 days of the life of the human embryo. The age of each embryo is estimated from what can be discovered of the time of ovulation. There is evidence that the viability of a mammalian ovum is measured only in hours and thus it is probable that estimations of embryonic age made in this way are fairly accurate. These first eleven horizons can be tabulated thus:

Horizon		
	I	One-cell fertilised egg.
	II	Segmenting egg.
	III	Free blastocyst.
	IV	Implanting ovum.
	V	Ovum implanted, but still avillous.
	VI	Primitive villi, distinct yolk sac.
	VII	Branching villi, axis of germ disc defined.
	VIII	Hensen's node, primitive groove.
	IX	Neural folds, elongated notochord.
	X–XI	Early somites present, up to 20.

Recovery of human ova and early embryonic stages is difficult and fortuitous. Few embryologists have even seen a human ovum or blastocyst; it is not surprising that when one is recovered it is described carefully.

Streeter's Horizons XII to XXIII cover the embryonic period during which the systems develop. It is the time of organogenesis: a chapter on each organ would not cover all that is known. The details are too recondite for this brief review, and we must be content to list only a few outwardly discernible features.

Horizon XII   3·2–3·8 mm, 26 ± 1 days old. Three pharyngeal bars present.

XIII   4·0–5·0 mm, 28 ± 1 days old. Arm and leg buds just formed.

XIV   6·0–7·0 mm, 28–30 days old. Leg buds fin-like, ear and eye defined.

XV   7·0–8·0 mm, 31·32 days old. Head large, nostrils forming, hand plate present.

XVI   8·0–11·0 mm, 33 ± 1 days old. Main parts of brain formed; thigh, leg and foot regions recognisable.

XVII   11·0–13·6 mm, 35 ± 1 days old. Head much larger, facial processes present; digital rays in hand.

XVIII   14·5–16·0 mm, 37 ± 1 days old. Body a more unified mass. Toe rays present; eyelid folds, tip of nose and ear hillocks discernible.

XIX   17·0–20·0 mm, 39 ± 1 days old. Trunk region starts to straighten out.

XX   21–23 mm, 41 ± 1 days old. Limbs increased in length and stubby fingers present.

XXI   24 mm, 43 ± 1 days old. Fingers longer, toes present.

XXII   25–27 mm, 45 ± 1 days old. Eyelids starting to cover eyeballs; ear assuming definitive form.

XXIII   28–30 mm, 47 ± 1 days old. Head bending into erect attitude; neck more apparent; limbs longer.

Eight weeks after fertilisation a human foetus is some 40 mm in C.R. length. It now has an unmistakably human appearance. The head is bulging and round, the forehead high, indicating an already relatively large brain. The face is flat, the nose small, but suggestions of a definite chin can be discerned. The ear has a shell-like pinna in miniature. The thorax is barrel-shaped and squat, the abdomen rotund and protuberant. The forelimb is more developed than the hind, but already certain characteristic features of fingers and toes are observable. The rudimentary tail has disappeared and there is a rounded rump. All these external features indicate that their owner must rank in a group high among Primates, but it is only its large brain that 'predicts that this being is destined to feel, think and strive beyond all other species that live on Earth'.

SUMMARY

Fertilisation of the ovum occurs in the outer third of the uterine tube. The human ovum is shed as a secondary oocyte, surrounded by its zona pellucida and corona radiata cells. Spermatozoa cause dissolution of the coronal cells, and while several may reach the zona and penetrate it, only one enters the oocyte, losing its middle piece and tail. The secondary oocyte undergoes its final maturation division, shedding a polar body, and its chromosomes form the female pronucleus which meets the male pronucleus formed from the head of the fertilising spermatozoon. Fertilisation restores the diploid number of chromosomes, determines sex and initiates cleavage. The resulting zygote continues to pass down the uterine tube; its single cell begins to divide, forming two, four, eight, sixteen, thirty-two cells still within the enclosing zona pellucida. It takes four to five days to reach the uterus, by which time the developing conceptus has become a solid ball of some 100 cells and has then become fluid-filled in the form of a sphere (a blastocyst) 0·09 to 0·1 mm in diameter. The investing single-celled layer is the trophoblast, destined to give rise to the chorion which in turn contributes to the placenta. Attached inside to a region of the trophoblast is a cluster of cells, the embryonic knot, which will become the embryo.

When a blastocyst reaches the uterus the lining endometrium has been stimulated by ovarian hormones to a stage of activity that ensures the subsequent existence of the conceptus. The zona pellucida is lost, and the blastocyst becomes attached to the endometrium and there follows the process of implantation, or nidation. The process differs in the various mammalian orders: in a human uterus the blastocyst becomes embedded within the endometrium by interstitial implantation. There is a mutual interaction between the tissues involved; several hormones and possibly other chemical substances are implicated in bringing about the cellular transformations. Some mammals exhibit a delay in implantation. The blastocyst remains quiescent for months, even a year, in the uterus until implantation is precipitated.

The implanted conceptus rapidly exhibits activity in its trophoblast, which becomes two-layered. An outer syncytial layer lacks cell membranes and makes intimate contact with the maternal tissue, now transformed into the decidua. An inner cellular layer is the source of the syncytium and also produces a gonadotrophin which is probably involved in suppressing the onset of the next menstrual loss. The trophoblast soon increases in complexity and develops villi that project into maternal blood spaces. Foetal blood vessels form in the villi and link up with those inside the embryo via a connecting stalk, thus establishing a foetal circulation through what will become the placenta.

The early embryonic knot of human blastocysts soon exhibits a cavity, formed by a hollowing-out process, with a floor of ectodermal embryonic cells. This amniotic cavity contains fluid and enlarges to surround the growing embryo, allowing it to move and also protecting it. A second, yolk sac, cavity forms below the embryonic disc; it is lined by endoderm. The ectoderm, endoderm and a third layer of cells, mesoderm, which develops between the other two, are the germ layers that give rise to all the tissues and organs of the embryo. The embryonic disc rapidly develops by processes of folding on itself and by the appearance of important structures within its substance. The notochord is an axial rod on each side of which segments or somites of mesoderm are formed. Two neural folds appear dorsal to it and fuse to form a neural tube from which the brain and spinal cord develop. As the yolk sac shrinks, its front end becomes enclosed by pharyngeal or branchial arches, homologous to gill bars in lower vertebrates. The remainder of the endoderm of the original yolk sac becomes the lining of the alimentary canal. The various stages in human embryonic development can be outlined in a succession of horizons during which morphogenetic changes become manifest one upon and after another. Eight weeks after fertilisation a human embryo is 40 mm from its crown to its rump. The main organ systems have formed, and it has an unmistakably human appearance.

6. Pregnancy, the Placenta and Birth

Pregnancy is without doubt a normal biological condition, but it occasions a progressive series of changes, both anatomical and physiological, which affect not only the generative system but the whole maternal organism. As pregnancy advances, there develop widespread alterations in many of the mother's organs, and the general metabolism of her body is profoundly modified. It is relevant to ask whether after the termination of one pregnancy, or of a succession, the mother is better off in her general condition, is unaffected by events, or in any way suffers irreparable damage. Are there any disadvantages in childbearing which leave an assembly of sub-standard matrons in the population? Or, on the other hand, does pregnancy result in the fulfilment and satisfaction of the female so that she develops her potential to her own advantage as well as that of her family?

Viviparity could not have occurred without the evolution of several important functions in maternal organs. There must first be the apparatus available for the protection, nutrition and provision to the growing embryo of all it needs during its intra-uterine existence. There must then be the means of its safe expulsion from the uterus, at the right time, without damage to the mother. Further, the maternal organism must be prepared for the subsequent care of the newborn young until it is large enough, strong enough and able enough to care for itself. Conversely, the neonate must be born capable of utilising the care-giving (epimeletic) attributes of its mother to satisfy its needs. Viviparity has resulted from the evolution of a whole series of adaptations that have become incorporated in complex interrelations of structure and function.

The human uterus undergoes a series of changes during pregnancy that are without parallel in any other organ. By 12 weeks it has become globular in shape and almost fills the pelvic cavity of the mother. By about 16 weeks it has become pear-shaped and its upper part can be felt through the mother's abdominal wall between the umbilicus and the pubes. At the end of 24 weeks it has distended enough for its upper level to be near the umbilicus. A few weeks before term the top of the uterus reaches as high as the level of the mother's lower rib margins at the front: often the level sinks a little during the last weeks. The increase in size keeps pace with that of the growing contents, until eventually the weight of the uterus is about 16 times that of the non-pregnant organ. There is a remarkable increase in number and size of the muscle fibres as pregnancy advances, and their power to stretch becomes more marked. The blood vessels also increase in size, becoming coiled and tortuous at first, but later becoming straightened. All through pregnancy the uterine muscle exhibits some activity. Intermittent contractions in early stages become more marked later. They are involuntary and are not appreciated by the mother, unlike the 'pains' at labour. The cervix becomes softer as pregnancy advances, due to increase in thickness of its inner lining and to its greater blood supply. The uterus is thus not only a muscular bag surrounding, protecting and providing nourishment to the foetus, but is also preparing for the expulsion of its contents at parturition. The arrangement of the muscle fibres also changes with expansion, so that the thick middle layer consists of numerous bundles spiralling about the organ, enhancing its expulsive power.

The skin and its derivatives, particularly the mammary glands, exhibit changes in pregnancy. The skin of the abdominal wall becomes stretched and pinkish, pearly lines appear there, on the breasts, thighs and buttocks (striae gravidarum), but there is much variation. After pregnancy the striae become silvery white: they may persist for a long time. Pigmentation also develops, to a varying degree. A dark vertical line appears on the middle of the abdominal skin, and patches of dark pigmentation may develop anywhere but more usually on the face. Sometimes the whole face is darkened, producing an almost mask-like effect called a chloasma ('greenness') of pregnancy. Most, sometimes all, of this pigmentation fades after parturition. Hair may be shed, there may be tiny red eruptions on the skin, the nails may break easily and skin glands may be more active. The breasts enlarge; in their periphery they may feel tense and even slightly tender to touch. Some clear secretion may be expressed from the nipple. The mammary tissue is responding to the hormonal climate generated as pregnancy advances. Pigmentation of the nipple and surrounding areola increases, more so in brunettes, and on the skin about the areola a less dark, mottled or almost ghostly web of pigmentation appears. On the areola itself a number of lightly coloured little tubercles become more prominent as a result of secretory activity (sebaceous). As these changes progress, striae may also develop. In women who have had children, many of these changes may already be present and a subsequent pregnancy merely enhances them.

The maternal circulatory system is obviously involved during pregnancy in maintaining an adequate blood supply to the body generally, and in particular to the uterus and the maternal side of the placenta. As pregnancy advances, the output of blood from the heart increases and the total blood volume in the maternal blood vessels also increases. The heart can usually meet the extra demands, and seldom becomes at all enlarged, but in doing so it calls on its reserves and therefore is not able so well to deal with any additional stress. The reserve power of the heart is in any event reduced because of the increased weight of the mother, the increase in the amount of blood and of the system of vessels containing the blood, and because of the embarrassment to the heart of the enlarged uterus pressing upwards against the diaphragm. There is often evidence of increased amounts of blood accumulating in the veins. In certain places in the skin and in the neck, the enlarged veins can be seen; there are often accompanying varicosities in leg veins and as haemorrhoids. The sluggishness of the flow of blood through veins is often evident in some swelling of the legs, also due to accumulation of fluid. These changes are to a reduced extent seen in many mammals late in pregnancy, but they are enhanced in human beings because of the adoption of an upright rather than a pronograde posture.

The maternal endocrine glands are profoundly influenced by pregnancy, both, as has been explained, in respect of the hormonal control of the reproductive organs and also as a result of demands by the growing foetus. The pituitary almost doubles its size, and in some mammals special pregnancy cells are discernible with differential stains. The thyroid frequently becomes enlarged in association with the increase in metabolic rate during the later months of pregnancy. The adrenal glands also enlarge, though their function is somewhat obscure: their increased activity may be a factor behind the pigmentation changes already noted. The maternal parathyroid glands enlarge in response to the extra demands for calcium for the foetal bones. There are marked changes therefore in the hormonal climate during pregnancy, and not unnaturally the foetus may be affected. Reactions by the foetus may be seen in a precocious stimulation of some of its own organs; these are most marked in foetal foals and seals. Human neonates exhibit such effects by the occasional ability to express a little watery secretion from their nipples (witch's milk), showing that maternal hormones have crossed the placenta to influence the foetal mammary tissue.

The maternal urinary tract, especially the ureters, becomes dilated during pregnancy. This is due partly to a fall in the tone of the muscles of the ureter and partly to a mechanical pressure exerted by the distending uterus. The dilatation persists throughout gestation and subsides some weeks after birth. The bladder may be irritated during early pregnancy and again near term, when micturition can be uncomfortable. The stomach, as is well known, may be reflexly disturbed in the early months, causing nausea and vomiting, and again towards the close of pregnancy there may be discomfort from the pressure below of the enlarged uterus. Reactions in the nervous system are variable, sometimes there being none at all, or, depending on the mother's temperament, expressions of an innate satisfaction or equanimity, to disturbances of an irritable nature, inability to sleep or, conversely, drowsiness. Vagaries of appetite, the so-called 'longings', and sudden compulsions to clean out the entire house are well known. Some women notice an enhanced awareness and sensitivity, a feeling of increased fitness especially in early months.

The foodstuffs supplied by the mother to the foetus normally do not come from reserves in her own tissues but from her own food supply. Only if the latter is insufficient are there calls on her tissues. The requirements of the embryo are at first inconsiderable, and it is known that the mother builds up larger stores of certain elements which are available not only as the foetus grows but later during lactation. Naturally there are increased demands for certain substances, protein, carbohydrate, calcium and iron; later in pregnancy there is a demand for fats as the foetus puts on a fatty layer beneath its skin. Again, only if her food is inadequate in these substances are her tissue reserves called on to supplement her dietary intake. Iron is in especial demand by all mammalian foetuses, and it is stored in the foetal liver as well as being utilised for making haemoglobin. There is virtually no iron in milk, and thus during the lactation period the iron is taken from the liver of the neonate to continue formation of haemoglobin.

The anatomical and physiological demands of normal pregnancy are, therefore, well within the capabilities of a healthy woman and other female mammals. The dramatic changes in certain organs can be easily accommodated as the result of adaptive changes, and the physiological response to the stimulus of pregnancy in many ways leaves the mother better off and prepared for the subsequent demands of lactation and care of her young. What disadvantages there are to the human manner of reproduction arise from three main sources: man's bipedalism; the internal relationships of foetus, placenta and uterus; the retarded development in certain aspects of a human foetus at birth. Some of these disadvantages will become apparent in the subsequent pages.

THE PLACENTA AND THE FOETAL MEMBRANES

The uterine tube and the uterine glands provide the first secretions nutritious to the cleaving egg and blastocyst, and the blastocyst obtains foodstuffs by diffusion as it invades the endometrium. From about the third week after fertilisation foetal blood vessels first make their appearance both inside the embryo and outside in the membranes surrounding it. The stage is now reached for the development of an organ, the placenta, that acts as an intermediary between the foetal and maternal circulations. The placenta shows a diversity of form greater than that of any other mammalian organ and is thus particularly difficult to define.

The term *placenta* (a Latin word meaning cake-like in form) was given to the discoid human afterbirth by the Italian anatomist Columbus about 400 years ago. In its present meaning it is applied to 'any intimate apposition of fusion of the foetal organs to the maternal (or paternal) tissues for physiological exchange'. This extended definition by an American embryologist, H. W. Mossman, allows us to include all connections, sometimes quite curious, between developing young and mother (or even father) in a wide variety of animals. It even enables us to consider relationships in certain invertebrates and lower vertebrates as well as those in highly specialised mammals.

The placenta has been considered a somewhat mysterious organ with a complex structure and not clearly understood functions. It is essential to grasp the ways in which certain foetal 'membranes' are transformed into the component parts of each type of placenta.

The human placenta, and that of the great apes, is discoid in shape; at birth it is flattened and circular like a plate. The human full-term placenta is nine inches in diameter, about an inch thick and weighs a pound. On its outer, or maternal aspect, there are 16–20 lobes or cotyledons, separated by clefts in which lie septa of maternal tissue. The cotyledons fit into corresponding depressions in the wall of the uterus. From the edge of the disc a membranous sac arises and surrounds

the foetus. Inspection shows that the sac has an outer layer
continuous with the substance of the placental disc, and inner
one (the *amnion*) that is reflected on to the umbilical cord. The
outer layer is the *chorion* (Greek for skin), a name first given
to it by Galen by analogy with the skin of a grape. It is
derived from the original outer covering of the blastocyst – the
trophoblast. Part of it has developed a large number of
processes, or *villi*, on its outer surface (chorion frondosum),
and it is these which become the cake-like placenta. In the

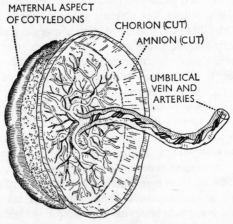

Fig. 14. *Diagram of a full-term human placenta after it has
been shed from the uterus. The chorion and amnion have
been cut. The umbilical vein conveys oxygenated blood from
the placenta to the foetus.*

villi foetal connective or mesodermal tissue forms the supporting tissue or core of the villi, and also gives rise to blood
vessels. In the thin part of the chorionic sac, attached to the
margins of the main disc, these two layers are fused and appear
almost as one. No villi are present, therefore this part is called
the chorion laeve (*laeve* means bald). Another way of illustra-
ting what has happened is to imagine that at implantation only
that part of the trophoblast near the uterine wall develops
villi, while that opposite is pushed out into the uterine lumen
as the embryo grows.

Other membranes, besides the chorion and amnion, take part in the formation of the mammalian placenta. In some species a large, sausage-shaped, fluid-filled sac is found within the chorion. This is the *allantois*. Initially it is an outgrowth from the hind end of the embryonic gut. The latter is at first a common chamber for lower ends of developing alimentary and urogenital systems. It is called a *cloaca* (Latin, meaning a sewer), and it later separates into the anal canal and the *urogenital sinus* which is destined to develop into the bladder and most of the urethra. The allantois retains connection with the urogenital sinus by a hollow tube that enters the embryo at the umbilicus. Remnants of its existence can be seen in the adult in the form of a strand of connective tissue, the *urachus*. It extends from the apex of the bladder to the umbilicus. The allantois is lined by endodermal cells similar to those lining the yolk sac. From the time it starts to grow out from the hind-gut it carries mesoderm with it on its outside. This allantoic mesoderm becomes vascularised, and if the allantois grows out far enough to reach the chorion its blood vessels become those of the placenta. We speak, therefore, of a *chorio-allantoic* placenta, and we mean by it *the* placenta – that is, the structure commonly called the afterbirth (Fig. 14).

The allantois is well developed in birds; it is vascular and essential for respiration, absorption, and excretion. Its form varies in marsupials, in which it may or may not reach the chorion to form a type of chorio-allantoic placenta. In Eutherian mammals it is occasionally very large, as in Carnivora, whales, lemurs, horses, ruminants, and may contain much fluid. The allantoic fluid is probably excreted by the foetus, either directly via the developing bladder or from the allantoic vessels. In Primates, including man, the endodermal allantois is small or absent, but as already indicated its mesoderm plays an important part in forming blood vessels.

The fourth foetal membrane in mammals is derived from the yolk sac. In birds it is developed by extension of endoderm about the yolk, with mesoderm following later and becoming vascular; it provides a means of mobilising yolk and conveying it as food to the embryo. In some mammals, despite the reduction of yolk, the endoderm of the yolk sac grows out and comes into contact with the chorion. This is a non-vascular yolk-sac placenta, and it occurs in many rodents, the armadillo and bats. There is fusion or intimate apposition of the bilaminar layer with uterine tissues, and physiological exchange occurs by diffusion.

In most marsupials, in mammals such as squirrels and insectivores mesoderm extends between the yolk-sac endoderm and the chorion. Vascularisation of mesoderm occurs, and blood circulates through the trilaminar structure to and from the embryo. This primitive type of yolk-sac placenta is the chorio-vitelline placenta. In the Eutheria it is very variable in size and importance, sometimes simple and bilaminar, sometimes trilaminar and vascular, and often transitory, being replaced by the chorio-allantoic placenta. In most rodents affairs are complicated in that the yolk sac becomes 'inverted' and eventually, owing to the breakdown of the bilaminar part, the vascularised endodermal part comes in direct communication with the uterine lumen. Recently it has been shown by experiment that this curious phenomenon may play a part in transfer of antibodies and perhaps food substances to the embryo. In Primates the history of the yolk sac is again variable, but a chorio-vitelline placenta cannot be formed. The yolk sac is largest in the marmosets. The small yolk sac in man is covered by vascularised mesoderm and is important as a site of blood formation in early embryonic life and also as the site of origin of germ cells.

Placental form

Both the form of the placenta, and the arrangement of villi when present, differ in the various groups of mammals. When fully developed, we can recognise the following forms. A *diffuse* placenta, in pigs, horses, and whales, possesses villi, or folds that resemble villi, distributed over the whole of the chorion. The villi may be restricted to limited areas called *cotyledons* (thus cotyledonary placenta); in some deer there are few cotyledons (5–8), whereas the goat and the cow have many (120–180). The villi fit into crypts in specialised portions of the uterus called caruncles, and thus a series of little placentae, or *placentomes*, are formed. These may look like flattened discs, oval plates, elongated ovoids or hollowed-out cups. In many Carnivora the placenta is *annular* or *zonary* and is like a 'muff' about the middle of the chorionic sac. In monkeys and in the tree shrew the placenta is bi-discoidal, with the two discs placed on opposite sides of the uterine cavity. Women occasionally have bi-discoidal, and rarely diffuse or annular placentae. In lemurs the placenta is diffuse and strikingly resembles that found in certain ungulates. Tarsiers, however, have a discoidal placenta which in many ways resembles that of man.

Placental form is correlated with the type of uterus, whether it be single or double horned, and whether there are caruncles or whether the endometrium is glandular throughout. It is also a reflection of the genetic constitution in each species, but is of little help in evolutionary classification. A discoid placenta can develop over the mouth of the uterus (the opening to the cervical canal). This condition is called *placenta praevia*, and in such a position the placenta may interfere with normal birth.

Placental villi exhibit many shapes. They are long and filiform in deer, and in reindeer each main stem villus has many side branches like a tree. In goats the villi branch frequently to form tufts. Branching may be so complicated in some mammals that a three-dimensional lattice-work is formed. The placental mass is thus sponge-like, and the arrangement of villi is labyrinthine. In some mammals the uterine lining remains intact and lies in close contact with the surface of the villi. In others there is a varying amount of destruction, and in yet others the maternal tissue is eroded until maternal blood escapes and percolates through the labyrinth of spaces between the villi.

The arrangement of the villi in man has been difficult to define precisely. It has been suggested that the villi are arranged in a digitate manner like fronds of seaweed, with maternal blood circulating through the intervillous lakes. Others have described a series of thick main-stem villi passing through the mass of the placenta, and from which many side branches are given off. These are turned back, so that the appearance is similar to that of many inverted willow trees. More recently, the view has developed that the villi may at times appear as described above but that tips of adjacent villi fuse together as they develop, with eventual formation of a sponge-like, *labyrinthine* arrangement.

The human placenta is *caducous* and *deciduate*. These terms imply that it is shed at term (*caducere* means to fall away) and that in its separation from the uterine wall some maternal tissue (*decidua*) is also torn away.

It might seem a strange waste that so useful an organ should be shed and discarded at the end of its life. Extrusion from the uterus occurs after the birth of the child; but separation starts early in labour, and contractions of the uterine muscle cause shearing forces in the placental bed. After expulsion, a raw, bleeding area is left at the placental site. Up to a pint of maternal blood may normally be lost from this site at the time of labour; eventually the vessels become occluded. There is a slight loss of blood and fluid for a few days of the *puerperium* (*puer* = a boy, and *parere* = to bear), and this constitutes the *lochia*. At first the lochia is red, but after about four days it becomes brown. By the tenth day it is yellowish-white or almost colourless, and consists principally of secretions from the cervical glands and cellular debris. The lochia ceases about the middle of the third week, but repair at the placental site is still incomplete. Shedding the placenta, and the damage it causes to maternal tissues, is potentially dangerous. This is particularly true when birth is assisted, because infection may be introduced. Micro-organisms are found in the human uterine cavity one day after birth, but fortunately most are harmless. In the past, when man had no knowledge of bacteria, pathogenic organisms frequently gained access to the raw surface of the placental site. They were literally carried to the female reproductive tract on the hands and clothing of the accoucheur. We little realise nowadays how devastating were the ravages of puerperal fever, and rightly we remember the name of Ignaz Philipp Semmelweiss (1818–65), a Hungarian physician who demonstrated the need for antisepsis during childbirth.

Many mammals shed their placenta, some even eat it and so return to the maternal organism much-needed substances such as the iron that is occasionally stored there. In other mammals the placenta is not shed (the contra-deciduate type) but degenerates and is absorbed by maternal tissues. In the non-deciduate mammals (pig, horse, whale, and also lemurs) the placenta is shed, but without any decidua. There are risks for the maternal mammal with all these types of placental fate, but to the human foetus with the deciduate type there are certain advantages. The placenta separates so easily that there is little chance of tearing of the cord and of blood loss from the foetus. Separation results in the placenta being squeezed by the contracting uterus and a quantity of blood is returned to the foetus before the cord is cut.

Fine structure

Variations in fine structure of the chorio-allantoic placenta, as seen through the light microscope, were used by a German placentologist, Otto Grosser, and later by an American embryologist, H. W. Mossman, as a means of classification. They examined 'the intimacy of contact between the chorion and the maternal tissues and thereby . . . the thickness and constitution of the membrane separating the maternal and foetal blood streams'. The two blood streams normally never mix, always being separated by placental and in some species maternal tissue.

Four main types of histological relationship have been recognised. It is assumed that the chorion is always an intact, continuous layer, however thin. The name of the type of maternal tissue found in apposition to the chorion is placed before the word chorion to describe each type of relationship. Thus *epitheliochorial* indicates that the maternal uterine epithelium lies in contact with chorion; such a relationship is found in the non-deciduate, diffuse type of placenta. Theoretically, every type of placenta commences by being epitheliochorial in that this is the relationship that pertains when trophoblast first comes into contact with maternal epithelium. There are no blood vessels in the chorion at this early stage, and so this is not an argument that an epitheliochorial type of placenta is necessarily primitive. The *syndesmochorial* relationship implies that maternal epithelium has been destroyed to a varying degree, and the chorion is thus in contact with maternal connective tissue. It is found in placentae of some ruminants, but is not as common as was once thought. In the *endotheliochorial* type the chorion eats through or destroys maternal epithelium and connective tissue until it comes into contact with the endothelial lining of maternal blood vessels, although connective tissue is often left or formed around each vessel. This arrangement is typical of Carnivora, some Insectivora, tree shrews and some bats. The fourth type is called *haemochorial*; the trophoblast destroys even the endothelial lining of the maternal blood vessels, and thus maternal blood escapes to bathe the outer surfaces of villi. A haemochorial placenta is found in some rodents, hedgehogs, monkeys, apes and man. Two varieties of haemochorial placenta are described: one is the labyrinthine type, in which villi are arranged in lamellae with maternal blood percolating through narrow capillary-like channels (as in rodents); the second is the villous type, in which villi float more freely in a pool of blood. The labyrinthine type is usually considered more primitive, and in Primates all variations are found between it and the villous type.

A distinguished embryologist, J. P. Hill (1873–1954), considered that four stages could be defined in Primate placentation: (a) a *Lemuroid* stage – in which the placenta is non-deciduate and epitheliochorial; (b) a *Tarsiiod* stage – discoid or localised, deciduate and haemochorial, with peculiar features distinguishing it from the placenta of monkeys and some characteristics of vascularisation not unlike those of the human placenta; (c) a *Pithecoid* stage, in which implantation is superficial and in which a bi-discoidal, labyrinthine, and haemochorial placenta develops; (d) an *Anthropoid* stage, with a deciduate, discoidal, villous, and haemochorial placenta. Man is, of course, included in the fourth stage, but many consider that the placenta is labyrinthine.

It is unfortunate that study of the placental structure has

thrown so little light on the evolution of mammals. Man displays some primitive features in placental development and structure, although there are also some specialised characteristics. Many consider that there are theoretical disadvantages associated with the haemochorial relationship, but man's reproductive rate suggests they cannot be very serious. There is evidence that circulation of blood through the intervillous space is slow, and that formation of localised clots of maternal blood is not uncommon. Maternal arteries leading through uterine tissue into the intervillous space are remarkably coiled, show curious changes in their inner linings, and have narrow nozzle-like openings; these anatomical features could lead to considerable slowing of blood flow. Blood leaves the intervillous space into numerous uterine veins opening near maternal septa that separate the main cotyledonary masses. The openings of these veins are wide enough to admit the tips of villi and may thus get blocked, leading to stagnation of blood flow. Small portions of the villi, as packages of trophoblast, are known to break off and pass into the maternal circulation. This has been considered a pathological invasion, but recently it has been suggested that it may have an important biological significance. The continued 'injection' of foetal tissue into the mother may be involved in a signalling system affecting her endocrine and immunological response to pregnancy.

Slowness of circulation rate in a haemochorial placenta may result in insufficient oxygen reaching the foetus. It is known that foetal haemoglobin has a marked affinity for oxygen and that other physiological mechanisms encourage the transfer of oxygen and may compensate for the slowness of the maternal blood flow. Yet it appears that the foetus lives in a state of relative oxygen lack, rather like that experienced by a man in climbing a high mountain. Some features of foetal blood, such as the frequently increased number of red cells (polycythaemia), their large size (macrocytosis) and the fact that it 'resembles in many respects . . . blood that has been subjected to an effective, continuous and extremely potent stimulus to blood formation' may be responses to an insufficient quantity of oxygen crossing the placenta. Therefore it could be argued that an intra-uterine existence any longer than necessary becomes increasingly dangerous as time passes. Prematurity and postmaturity are both undesirable to a foetus.

Placental transfer

The placenta performs for the foetus the functions of respiration, food absorption and excretion. It has a large surface area in contact with maternal tissue or blood to carry out these functions. The villi of a full-term human placenta have a surface area of about ten square metres. The cells covering the villi possess minute processes on their surfaces called *microvilli*. They can just be seen with the light microscope as a 'brush border' to the chorionic cells. The electron microscope demonstrates them in the form of slender protoplasmic strands, often with dilated ends.

Many substances pass unaltered through the placental membrane; some, such as iron, glycogen and fat, are stored, others are modified during transmission, and others are aided across the membrane. Many substances never normally pass across. Oxygen and carbon dioxide pass in either direction by diffusion under pressure, but efficiency of oxygen transfer is about one-twentieth of that across the lung. Sugars, lipoids and proteins do not cross by simple diffusion, and various enzyme systems may be involved. Amino acids can cross the placental membrane slowly, even when the concentration in foetal blood is higher than in maternal blood. Transfer against the gradient can also occur with other substances.

Certain poisons and drugs pass easily across the placenta, including anaesthetics and alcohol. Some can affect the developing organs of the foetus, or alter their activities. Great care should be exercised in giving drugs during pregnancy, as the thalidomide tragedies have only too clearly indicated. Much research has been done with radioactive isotopes of those elements that normally traverse the placenta. It is generally too dangerous to experiment with such substances during human pregnancy, and so such work is most frequently carried out on experimental animals. It is also possible to 'label' maternal red cells with radioactive isotopes and, after injecting them back into the maternal circulation, to find out if there is any mixing of the labelled cells with foetal ones. Normally no mixing occurs, but in certain pathological conditions there is some evidence that it does.

The placenta acts as a most effective barrier against transmission of bacteria. Organisms are believed to be able to pass to the foetus only after producing a lesion in the placenta. Some viruses (that of German measles, *rubella*), or toxic substances elaborated by them, can cross the barrier and may interfere with developmental processes. Proteins in immunological quantities can cross as unaltered molecules, as can many antibodies. The Rh agglutinogen passes from foetal to maternal side, and the subsequently formed antibody (isoagglutinin) can pass back to the foetus. It may return in quantities sufficient to cause much destruction of foetal blood. It is for this reason that no Rh negative woman should be transfused with Rh positive blood. In at least 10 per cent of all pregnancies conditions are such that iso-immunisation by Rh factor could occur. Blood-destroying disease in all its forms, however, only complicates 0·25 per cent of pregnancies and even more rarely is it very severe.

Both layers of trophoblast produce hormones: the placenta is therefore an endocrine gland. The human cytotrophoblast is believed to be the source of probably more than one gonadotrophin. It may produce slightly different gonadotrophins at different stages of pregnancy.

The human syncytiotrophoblast secretes oestrogens and progestins. To what extent placentae of other mammals secrete hormones is not fully known. The placental hormones pass into the maternal circulation and influence maternal endocrine organs; they may play some part in controlling the onset of labour. If so, the placenta has another function in that it influences the length of gestation.

The placenta also acts as a barrier against maternal hormones, although some can cross and cause changes in foetal organs. Madame Dantschakoff has vividly discussed the importance of the placenta in preventing male foetuses being dangerously affected by maternal oestrogens: 'Truly, nature has taken especial pains over the male sex.'

GROWTH OF THE FOETUS

The term foetus is applied to an embryo when it has acquired all the characteristics that can be recognised in later life. This stage is reached by the end of the second month in human embryos. The foetal period is thus mainly one of growth with certain alterations in shape and proportion. Some organs, the nervous system and the special senses do, however, continue to differentiate during the foetal period, and the brain does so even after birth.

Fusion of eyelids over the eyeball occurs during the third month. Most of the primary ossification centres appear in bones, and the first signs of hair and nails are seen. The face assumes a more human appearance during the fourth and fifth months; the skin glands become active. The foetus can now move powerfully enough for the mother to appreciate 'quickening'. Eyebrows are present, and eyelids are no longer fused by the seventh month; head hair is present. Children born prematurely at this time (or even a little earlier) can survive with careful nursing.

A full-term child weighs on the average $7\frac{1}{2}$ lb (3400 g), but the weight can vary from 6 lb (2700 g) to over 10 lb (4500 g). Length from crown of head to heel is 20–21 inches (50–53 cm), and C.R. length is 12–13 inches (30–33 cm). Circumference of the head is about 13 inches (33 cm).

A human neonate has a 'chubby' appearance, and its head appears unduly large relative to the body. The head is often distorted in shape, having a 'sugar-loaf' form because of moulding during passage through the birth canal. Moulding is quite normal and persists for less than a week. A region soft to the touch and devoid of underlying bone can be felt on the top of the head just above the hair-line. It marks the anterior fontanelle of the skull, a diamond-shaped area where there is membrane instead of bone. The bone grows into and obliterates the fontanelle when the baby is nine months old. The face is broad, and cheeks bulge owing to a pad of fat in each that facilitates sucking at the breast. The nose is broad and squat, and the facial region is relatively small compared with the rest of the large head. Jaws are small, devoid of erupted teeth (except, rarely, lower central incisors), and a distinct chin can be felt. The neck is apparently short, and there are marked circular creases in its skin. The bones of the neck (cervical vertebrae) are, however, relatively long at birth. Shoulders are set high on the thorax as if in a fixed shrug. This, and the fat neck and a tendency for the head to fall forward, give the short-necked appearance.

The thorax is more barrel-shaped, the heart relatively larger and higher within it, than in the adult. The abdomen is plump and distended owing to the relatively large liver – it occupies about half the abdominal cavity – and because marked elongation due to occur later in the lumbar spine has not begun. The 'small of the back' and the loins are therefore hardly discernible. The liver is so large because it is an important site of red cell formation in a foetus; this function is taken over by the bone marrow as the bones enlarge. Arms and legs are about equal in length; the leg steadily elongates from birth onwards. Testes are usually in the scrotum but may not have descended from the inguinal canal.

The human neonate is born with hair on its head and often some hair or a few single stiff hairs on other parts of its body. This is primary hair, *lanugo*; it is shed during the few weeks following birth and is replaced by secondary hair that may not be of similar colour. Lanugo is usually first shed from the back of the head, rubbed away by nodding movements against a pillow. A newborn child is covered to varying degrees by a white cheese-like substance of fatty consistency, *vernix caseosa*; it is washed off at the first bath. It is composed of dead skin cells mixed with sebum and is said to protect foetal skin from amniotic fluid. Vernix may break loose from the skin and be swallowed, sometimes with shed lanugo hairs, when a foetus drinks amniotic fluid. Finger nails reach the level of the tip of the pulp of the fingers by the end of a full-term pregnancy. All neonates are blue-eyed; other eye colours develop by six months as pigment is laid down in the iris.

PARTURITION

Pregnancy terminates at parturition, in the birth of the young. In women, labour is the process by which a foetus of viable age is expelled from the uterus. Normal labour implies a natural termination, with the foetus presenting by the vertex (the crown of the head), without artificial aid and without complications. It is usually divided into three stages. During the preparatory first stage dilatation occurs of the lower uterine segment and of the cervix. As dilatation progresses, the upper part of the cervix becomes merged into the lower uterine segment and eventually more of the cervix is 'taken up' into the uterus. This stage lasts about 16 hours in primigravida and about eight in multipara.

That portion of the foetal membranes pressing against the cervical region usually ruptures during the first stage with the escape of some amniotic fluid (the 'fore-waters'). The foetus will escape from the investing membranes by passing through the tear in them opposite the cervical opening. Sometimes the membranes do not rupture and the conceptus is expelled entire with the foetus enclosed in the chorion (born in a caul).

The second stage begins when dilatation of the cervix is complete and ends with complete expulsion of the foetus from the birth canal. The uterus, cervix and vagina become merged into a single broad channel through which the head and body of the foetus gradually passes. After the escape of the foetus there is a sharp loss of some blood-stained amniotic fluid. The stage lasts about two hours in primigravid women but only about ten minutes in multipara, depending on the power of the uterus. The uterus now shrinks in size, contracting intermittently, and the placenta becomes completely separated from the uterine wall.

The third stage of labour is marked by the expulsion of the placenta, about ten to fifteen minutes after the birth of the foetus, and it is usually accomplished by voluntary efforts on the part of the mother. As separation of the placenta is accompanied by haemorrhage from the uterine wall and placenta (see haemochorial placentation above), the delivery of the placenta is followed by the extrusion of blood and blood-clot. After labour the uterus is small, firm and hard but occasional intermittent contractions can be detected.

The way in which the foetus is arranged in the uterus before birth is variable. It may be described first in the way the foetus lies relative to the long axis of the uterus; longitudinal, transverse or oblique. Then it depends whether the crown of the head (vertex) or the buttocks (breech) 'present' first at the birth canal. Further, with the vertex presenting, the occiput of the foetal head may point in one of four main directions; to

CORD   PLACENTA

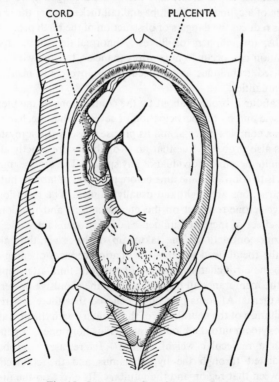

Fig. 15. *Diagram to show the arrangement of a foetus and placenta within a pregnant uterus. The foetus is lying with its vertex presenting.*

the front, left or right; to the back, left or right; each position is, of course, relative to the mother's front or back. The commonest position (50%–60%) in a vertex presentation is with the foetal occiput pointing to the front and to the mother's left. The lie and position of the foetus are ordained partly by the shape of the unicornuate uterus and partly by the way the foetus has its limbs, head and trunk 'packed' into the smallest possible space: i.e. all its parts are flexed into a compact ovoid. The disposition of the mother's organs within her abdomen also to some extent affects the lie. Foetuses of mammals with a bicornuate uterus are arranged more in the form of a cylinder, with limbs and tail tucked under the trunk. This reduces the dangers of obstruction of the birth process by limbs, fins, flippers or flukes. Abnormalities in the foetal position can occur in human pregnancies, but midwifery has devised techniques and procedures to overcome nearly all eventualities.

Labour is brought about by the contractions of the uterine muscle and by the participation of several other mechanisms. It is a complex process, and its precise cause is still a mystery. It is also probably true that no two labours are exactly alike, even in the same individual of any species. The uterine muscle starts to contract some time before the onset of labour, but the contractions are slight and usually painless. The recession of pregesterone restraint on the uterine muscle and the dominant effect of increased influences of oestrogens intensifies the uterine contractions. The 'taking up' of the cervix, the sinking of the foetal head towards the dilating cervical opening and into the birth channel, the loss of amniotic fluid after rupture of the membranes all help to increase the expulsive powers of the uterus. At some point it is believed that there is a reflex excitation of the neural lobe of the pituitary with the liberation of oxytocin into the blood stream. This hormone is a peptide with a molecular weight of 1007. Its release is probably mediated through the hypothalamus and the connections between that region and the pituitary. Its effect on the uterus is to increase the uterine contractions, and prior to parturition the uterine muscle becomes particularly sensitive to low concentrations of oxytocin. The reasons for this enhanced sensitivity are not clear, but it does appear that it can vary almost from hour to hour during parturition. As soon as the second stage of labour commences, reflex abdominal contractions develop in man and most mammals. These reflex contractions in the maternal abdominal muscle are brought about by distension of the vagina with the advance of the foetus.

Other, more mechanical, factors aid parturition. There is a relaxation of the maternal tissues, of the ligaments of the pelvic bones. In some mammals this is aided by a nonsteroid hormone, relaxin, probably produced by the ovary and perhaps by the placenta. Relaxin has a marked effect in guinea-pigs, softening and causing absorption of the pubic symphysis, and allowing birth of a remarkably large and well-developed foetus. The hormone may also permit a sow to give birth while asleep to twelve young. It is not known how much is produced in man, but relaxin from other sources has been used to aid softening of the human cervix.

There is also some moulding of the foetus as it passes through the birth channel, as well as a variable degree of rotation of the foetal parts. There is at first an attitude of flexion in the foetus, encouraged by the uterine contractions. This is succeeded by movements of extension as the head emerges. The degree of moulding is proportional to the pressure to which the foetus, especially its head, is subjected during labour. A natural degree of moulding is harmless to the foetus and is obviously advantageous. Should the head be oversized, or the maternal pelvis undersized, there may then be danger of damage to the foetus. Such possibilities can be determined antenatally and appropriate measures devised for successful delivery, perhaps by Caesarian section. Not infrequently a swelling can be seen on the top of the head of a newborn. It is known as a caput succedaneum and shows where uterine pressure forced the flexed foetal head against the partially dilated maternal tissues. These form a girdle of contact about the foetal head, and the central scalp tissue becomes swollen because of accumulation of fluid. Evidence of moulding and the caput normally subside with a day or two of birth.

The mechanism of normal labour, the apparently complex internal rotations of the human foetus are associated with the anatomical characteristics of man, both as adult and as foetus. The shape of the uterus, the anatomy of the human pelvis, the large size of the foetal head at birth, are primary factors ordaining the mechanism by which a foetus is expelled. It is interesting to ask whether man would be now so prolific if he had not so effectively mastered the risks and dangers attending childbirth. These not only include those of the process of birth but of infection and child care as well. The population surge began before midwifery reached its modern perfection: other more potent factors affect natality.

LACTATION AND MILK

The production of milk can properly only occur when the mammary glands have reached a certain state of development. Much of the growth of the gland tissue occurs during pregnancy; it starts as early as the second month. Growth of the mammary duct system and the secretory elements is brought about by the action of ovarian and placental steroid hormones (oestrogens and progesterone) together with a direct synergistic effect of anterior pituitary hormones. These include luteotrophin (prolactin) and the growth promoting hormone (somatotrophin, STH) as indicated in Fig. 6 (p. 45). There have been suggestions that specific 'mammogens' were produced by the anterior pituitary, but there are now doubts as to the existence of such substances. There is evidence that other hormones, from the adrenal and the placenta in some species, may exert a mammogenic action.

Lactation is the term used to include both milk *secretion* and milk *removal* from the mammary gland. Secretory activity starts during pregnancy, but is inhibited by the combination of oestrogens and progesterone then prevailing. At parturition the fall in the levels of these two hormones removes this inhibitory effect and leaves the mammary gland receptive to stimulation by lactogens of pituitary origin. There is still believed to be sufficient oestrogen present to exert its own lactogenic action and also to increase secretion of the pituitary lactogens. These appear not to be one single substance, but a complex including prolactin (luteotrophin) and adrenocorticotrophin (ACTH). The latter, as well as other hormones, play a part in lactation by regulating water and salt balance, sugar and fat metabolism, all of which are important in affecting the volume and composition of milk. It seems likely that the precise mechanisms involved in initiating abundant milk secretion vary in each species of mammal.

Milk removal, or ejection, is influenced by the production of oxytocin from the posterior pituitary. It has long been known that suckling exerts a stimulating effect on lactation. Stimulation of the nipple at suckling causes excitation through reflex nervous pathways of the hypothalamus and thus the pituitary. Output of both lactogenic hormone and oxytocin occurs, one affects milk secretion, and the other, milk ejection. The milk yield is thus the combination of both processes. The alveoli of the mammary gland and its terminal ductules possess contractile myoepithelial cells which help to 'let down' the milk into the main ducts and sometimes bring about a spontaneous emission of part of it. The more the milk is removed, the more secretion is this stimulated by local and hormonal action. Once established, mammary activity is maintained for many months. It is an essential sequential event to uterine gestation in that lactation provides the young with nutrition while it continues development and also supplies it with a store of foodstuffs to tide it over a weaning period that leads to eventual independence.

Milk is not produced until a day or two after parturition. The first secretion is of a watery nature, called colostrum; it may be produced even before birth. It contains numbers of cells and more protein than in later secretions. An intermediate type of milk is produced during the first month, followed by mature milk.

From Table 6 it will be seen that there is some variation in the composition of milk of different species. Cow's milk contains more protein than human milk, and when given to babies forms indigestible curds in the stomach. The protein may be removed by boiling the milk. There is much the same

Table 6. *Composition of milk (in grammes per cent).*

	Protein	Lactose	Fat	Calcium
Human colostrum	8·5	3·5	2·5	
Human mature milk	1·5	6·5+	3·0+	0·03
Cow's milk	3·5	4·5	3·5	0·15
Whale milk	10·0	1·0	50·0	0·3

quantity of fat in cow's and human milk, but it has a different chemical composition. Milk of seals and whales has a greatly increased amount of fat, and the pups and calves put on weight rapidly. A seal pup doubles its birth weight in a week, laying down much blubber. A nursing mother can produce 1·5 to 2·0 litres of milk a day, but a human baby takes about 180 days to double its birth weight. A large whale is estimated to produce 600 litres of milk a day!

The length of the lactation period also varies in different mammalian species. It is not related so much to the size of the mother or the young at birth as to the degree of development of the neonate. The suckling period is about the same length, three weeks, in rats and mice with small poorly developed young as it is for large seal pups which are precociously developed. Large whales lactate for about a year, elephants even longer. Human females often nurse for up to a year and can do so for longer. After about nine months the infant needs more food and, in particular, substances such as iron, than the mother's milk can provide. Weaning then becomes more urgent and eventually inevitable.

In human females the production of lactogens from the anterior pituitary inhibits the reproductive cycle during the nursing period. Usually, but not always, ovulation is inhibited and menstruation suppressed. The ovarian cycle is re-established as lactation ceases. Many mammals, however, exhibit a post-parturient ovulation some days or weeks after parturition. The female is, therefore, lactating and pregnant at the same time, or, more rarely is lactating and exhibiting either delay in implantation due to lactation or due to more complex factors (see p. 84). A post-parturient ovulation may thus cause a 'compression' of reproductive events that increases fertility over a period of time. It would be an advantageous factor in under-population or where adverse influences cause high mortality of young. It is fortunate that it is only rarely exhibited by human females.

SUMMARY

Pregnancy is associated with a series of profound changes in the reproductive organs as well as in the whole maternal organism. Not only is there a dramatic enlargement of the uterus until it is 16 times its original weight, but there are also anatomical and physiological alterations in most of the systems of the maternal body. These are particularly pronounced in women and may be correlated with man's upright posture, with the conditions in the human uterus and also with the demands of the growing foetus.

There are characteristic changes in the skin, with striae and pigmentation appearing in certain regions. The breasts enlarge, and changes occur in the areola. Maternal endocrine glands are all affected in association with the hormonal climatic responses to gestation. There is an increase in blood volume and body fluids, and thus there are increased calls on the power of the maternal heart. There are also certain reflex manifestations; most are attributable to the changes in the uterus. Foodstuffs are normally supplied to the foetus from the mother's food supply and not from her reserves. The mother builds up stores of certain elements for demands late in pregnancy and in preparation for lactation. The demands of pregnancy are well within the capabilities of a healthy woman, and at the end of pregnancy she is generally physically better off in many respects than at its start.

Nutrition of the young conceptus is at first obtained by diffusion. Later, certain foetal membranes are developed which in various ways, depending on the species of mammal, become transformed into a placenta. The membranes are: the amnion, surrounding the embryo; the chorion, containing the entire conceptus; the yolk sac and the allantois, both derived from the endodermal component of the embryo. As development proceeds, the embryo obtains all the parts characteristic of its species and thereafter it is called a foetus. This occurs by the end of the second month in human embryos, and thereafter there are mainly changes of growth and alteration in shape and proportion. Early in the foetal period the placenta develops its final form and characteristics.

The human placenta may be described by applying to it certain adjectives. It is usually discoidal in shape and applied to the upper posterior part of the uterus. It is caducous in that it is shed at the end of pregnancy, and deciduate in that some maternal tissue from the uterine wall is shed with the foetal elements. It is composed of numerous villi containing foetal blood vessels. The villi are arranged in a labyrinthine manner, with maternal blood percolating through the intervillous spaces. The maternal blood is in contact with the covering tissue of a villus, and its layers only separate the mother's blood from that of the foetus. This relationship is termed haemochorial. There are both advantages and disadvantages to this type of placentation.

The placental tissue separating the maternal and foetal blood streams has the functions of transporting water, gases, nutriments and other essentials to the foetus. It also acts as an excretory organ, transferring foetal waste products to the maternal circulation. Some substances cross relatively easily from one circulation to the other, some with difficulty and some normally not at all. There is a 'barrier' effect exerted by the placental tissue, and it is important for the foetus in that it prevents the passage of substances and micro-organisms that might be harmful. Some drugs, toxins and chemical substances can cross the placental barrier, especially in animals, and can interfere with embryonic development; they are known as teratogens if they cause abnormalities.

Parturition is the birth of a foetus. Three stages are usually recognised in human labour. The first is preparatory in that dilatation of the cervix occurs to allow expulsion of the foetus. The second stage involves the expulsion of the foetus through the birth canal to the exterior, and during the third the placenta separates from the uterine wall and is expelled. The mechanism of labour depends on the arrangement (position and presentation) of the foetus at the start of parturition. The causes of labour are complex and are partly hormonal and partly physical.

The human neonate is poorly equipped for an independent existence: there is a long lactation period. Despite a long gestation period the nervous system still needs several years for further development and differentiation. The child retains some foetal characteristics for longer than any other mammalian young, and childhood is more prolonged.

Synopsis

World population is increasing at such a rate that it will probably double by the end of the century. Not only is there danger of the population being inadequately fed, of famine and starvation, but there are also other drawbacks to a population avalanche. These include impoverishment of the environment, overcrowding and lack of privacy, reduction in opportunities for success, and increasing regimentation. Control of reproduction demands a full knowledge of reproductive patterns and organs, and of the factors that influence them. It also demands that full consideration be given to the possible dangers of short- and long-term interference in reproductive events for purposes of fertility control.

There are several main aspects of reproductive processes and patterns which require detailed investigation. The origins and methods of formation of the germ cells (ova and spermatozoa) must be determined. All the factors concerned in germ cell *maturation* in the gonads and related duct systems must be understood. This also involves the length of life of germ cells after their release from the gonads. The essential importance of germ cells is that they are transported so that spermatozoa may effect a *union* with ova at fertilisation. This marks the establishment of pregnancy, during which the *maintenance* of normal development in an embryo must be procured. There follows the successful *termination* of pregnancy at birth of the young, succeeded by a period of their immediate *survival.*

The reproductive pattern of any mammal is controlled through the output of gonadotrophic hormones by the anterior pituitary, itself under the influence of the hypothalamus. The hormones have actions on ovarian elements, causing follicular

126

growth, ovulation and the formation of a corpus luteum. The essential biological event in any pattern is ovulation, the liberation of an oocyte at the right time in the pattern. Before and after ovulation, steroid hormones from ovarian follicles and corpora lutea prepare the lining of the uterus for reception (implantation) of the fertilised egg, which by the time it reaches the uterus has developed into a blastocyst.

Germ cells migrate from the yolk sac of mammalian embryos to the gonads. All the oocytes are formed in the ovary by the time of birth or just after. On the other hand, spermatogenesis continues in the seminiferous tubules from puberty onwards. Oocytes are much larger than spermatozoa, but lack their mobility and must be transported by active movements of the female ducts. Ova are viable for a much shorter period than spermatozoa, and fertilisation must occur close to the time of ovulation. The various contraceptive methods devised for control of fertility endeavour either to prevent the union of germ cells or to inhibit the liberation of an oocyte at ovulation.

Fertilisation initiates the changes that will lead eventually to the development of a new individual. Sex is determined at fertilisation. Early divisions in the zygote form cleavage stages, leading to a morula and a blastocyst. Implantation denotes the attachment of the blastocyst to the uterine lining and its subsequent nidation. The trophoblast develops into a placenta, an organ in apposition to maternal tissue that carries out physiological exchange.

Intra-uterine life is characterised by particular adaptations, both anatomical and physiological, and in both mother and foetus. A certain hormonal climate is also necessary for successful gestation, and the human placenta plays an important role in prolonging and possibly in terminating gestation. The conceptus passes first through an embryonic phase, during which main organ systems develop, followed by a foetal phase marked by growth changes. Mammalian foetuses vary in their degree of development at birth; a human foetus is strikingly unprepared for independent existence and requires prolonged maternal care. Adaptations for viviparity include development of mammary glands and an adequate lactation period.

Bibliography

Books, reference works and research papers on reproductive anatomy, physiology and endocrinology are now so numerous that a classified monthly title list (*Bibliography of reproduction*) is now published by the Reproduction Research Information Service Ltd., 8 Jesus Lane, Cambridge.

The following is a list of books, by no means complete, which will give the reader a wide introduction to the literature.

1. CORNER, G.W. 1963. *The hormones in human reproduction.* Atheneum, New York.
2. BULLOUGH, W.S. 1961. *Vertebrate reproductive cycles.* Methuen, London.
3. VELARDO, J T. 1958. *Essentials of human reproduction.* Oxford Univ. Press, London.
4. ASDELL, S.A. 1964. *Patterns of mammalian reproduction.* Constable, London
5. HAMILTON, W.J., BOYD, J.D., and MOSSMAN, H.W. 1962. *Human embryology.* Heffer, Cambridge.
6. DICKINSON, R.I. 1962. *Atlas of human sex anatomy.* Williams & Wilkins, Baltimore, Md.
7. WOLSTENHOLME, G.E.W. (ed.). 1953. *Mammalian germ cells.* Churchill, London.
8. AUSTIN, C.R. 1961. *The mammalian egg.* Charles C. Thomas, Springfield, Ill.
9. ZUCKERMAN, S. (ed.). 1962. *The ovary.* Academic Press, New York.
10. PINCUS, G., THIMANN, K.V., and (in Vol. IV) ASTWOOD, E.B. 1948–64. *The hormones.* Academic Press, New York.
11. VILLEE, C.A. (ed.). 1961. *The control of ovulation.* Pergamon Press, Oxford.
12. HARTMAN, C.G. (ed.). 1963. *Mechanisms concerned with conception.* Macmillan, London.
13. PINCUS, G. 1965. *The conrol of fertility.* Academic Press, New York.
14. VILLEE, C.A. (ed.). 1960. *The placenta and fetal membranes.* Williams & Wilkins, Baltimore Md.
15. FLEXNER, L.B., (then) VILLEE, C.A. (ed.). 1954 onwards. *Gestation conferences.* Josiah Macy, Jr. Foundation, New York.
16. ENDERS, A.C. (ed.). 1963. *Delayed implantation.* Univ. of Chicago Press, Chicago.
17. ROWLANDS, I.W. (ed.). 1966. *Comparative biology of reproduction in mammals.* Academic Press, New York.

18. Young, W.C. (ed.). 1961. *Sex and internal secret'ons*. Williams & Wilkins, Baltimore, Md.
19. Parkes, A.S. (ed.). 1966. *Marshall's physiology of reproduction*, 3rd ed. Longmans, Green London.

NOTES ON THESE BOOKS

1, 2, 3. General descriptions of reproductive events, with some detailed information.
4. An exhaustive catalogue of patterns of reproduction in all mammals for which information is available; an invaluable reference work.
5. A textbook for medical students; mostly embryology.
6. An atlas especially useful for teachers.
7, 8. Specialists' reports on germ cells.
9. An exhaustive multi-author work on all aspects of ovarian biology.
10. A four-volume review of all hormones – for experts.
11, 12, 13. Excellent books for specialists covering research on ovulation, conception and fertility.
14. Useful review volume on the placenta.
15. Series of chapters by experts on many aspects of reproduction, especially gestation.
16, 17. Collection of reports on implantation, delayed implantation, and reproduction in some less well-known forms.
18, 19. Multivolume reviews by experts, with extensive bibliographies. These are essential reference works for all students and research workers interested in reproductive physiology.

Index